Perspectives on Digital Humanism

Hannes Werthner • Erich Prem • Edward A. Lee •
Carlo Ghezzi

Editors

Perspectives on Digital Humanism

 Springer

Editors
Hannes Werthner
Vienna University of Technology
Vienna, Austria

Erich Prem
eutema GmbH and Vienna University
of Technology
Vienna, Austria

Edward A. Lee
University of California
Berkeley, CA, USA

Carlo Ghezzi
Politecnico di Milano
Milano, Italy

ISBN 978-3-030-86146-9   ISBN 978-3-030-86144-5   (eBook)
https://doi.org/10.1007/978-3-030-86144-5

This Springer imprint is published by the registered company Springer Nature Switzerland AG.
The registered company address is: Gewerbestrasse 11, 6330 Cham, Switzerland

Preface

"This is absolute nonsense." This was the reaction of the audience, both academics and non-academics, participating at the First International Conference on IT and Tourism, in Innsbruck in 1994. Beat Schmid (University of St. Gallen, Switzerland) spoke about electronic markets and Larry Press (UCLA, USA) about digital agents.

Now, only 28 years later, this "nonsense" runs the world, Information Technology and its artifacts act as the operating system of our life, and it is hard to distinguish between the real and the virtual. We cannot imagine a world without it, and—besides running the world—it contributes and will continue to contribute to solving important problems. However, this comes also with interconnected shortcomings, and in some cases, it even puts into question the sovereignty of states. Other critical problems are echo chambers and fake news, the questioned role of humans in AI and decision-making, the increasingly pressing privacy concerns, and the future of work.

This "double face" is why we started the Digital Humanism initiative, with a first workshop in April 2019 in Vienna. Over 100 attendees from academia, government, industry, and civil society participated in this lively 2-day workshop. We talked about technical, political, economic, societal, and legal issues, benefiting from contributions from different disciplines such as political science, law, sociology, history, anthropology, philosophy, economics, and informatics. At the center of the discussion was the relationship between computer science/informatics and society, or, as expressed during the workshop, the co-evolution of information technology and humankind. The major outcome was the *Vienna Manifesto on Digital Humanism*, now available in seven languages, which lays down the core principles of our initiative.

Since then, we have organized a set of workshops and panel discussions. These events, forced by the pandemic to be online, have drawn a growing worldwide community. In addition, we succeeded in establishing a core group of internationally renowned intellectuals from different disciplines, forming a program committee that jointly "governs" and directs the Digital Humanism initiative.

We share the vision that we need to analyze and to reflect on the relationship between human and machine, and, equally important, to influence its development for better living and society. Technology is for people and not the other way around. We chose the term Digital Humanism, which was introduced–in the German-speaking world–by Julian Nida-Rümelin and Nathalie Weidenfeld with their book *Digitaler Humanismus* (Piper Verlag, 2018). We want to stress that humans should be at the center of the digital world. Technological progress should improve human freedom, peace, and progress in harmony with nature.

Today, the spirit of humanism should inspire the developments of our society, which is largely reliant on digital technologies. As such, we distinguish Digital Humanism from the Digital Humanities, the study of human society and culture pursued by digital means. In contrast, Digital Humanism aims at rethinking our current digital practices, including research, development, and innovation in the digital domain. As such, it maintains a positivist goal for technology to create societal progress rather than just innovation for the sake of economic growth.

The term *humanism*, taking a historical perspective, refers to two rather different movements. The first denotes the period between the mid-fifteenth until the end of the sixteenth century (Renaissance Humanism), with a rediscovery of antiquity in the arts and in philosophy, and in which scholars, philosophers, and artists were called and called themselves "humanists." Aesthetics and ethics became centered on humans rather than on the supernatural or divine. The best-known iconic represen-tation of Humanism is Leonardo da Vinci's *Vitruvian Man*, where a human enclosed in a circle is shown as the archetype of the principles of harmony and proportion advocated in Vitruvius' book *De Architectura*. A second period of humanism flourished in the Enlightenment period (end of eighteenth century), and the French revolution was largely inspired by the principles of human freedom and democracy rooted in the humanistic spirit of that time. Humanism was associated with educa-tional and pedagogical ideals that focused on values such as human dignity and humanity. Naturally, the two movements share a range of common concepts and interests, some of which remain relevant for Digital Humanism today, for example, a strong focus on human rights and how to maintain them in the digital realm.

There are, however, critics of these classical notions of humanism. Especially, the educational ideal of humanists has been criticized as supporting beliefs in European cultural supremacy. Furthermore, a focus on the human subject always requires critical examination regarding who that subject precisely is and which of its many traits should be considered essential. However, Digital Humanism today certainly has no supremacy or colonial mission; quite the contrary, it is critical of already existing colonial tendencies in today's digital technologies. This is evidenced by our stance on digital sovereignty and geopolitics, for example. Similarly, the question of which traits of human nature should be focused on is a subject of discussion in Digital Humanism, especially since the relation of the individual and the society is a major concern of digital humanists.

In the context of the Enlightenment, proponents of Digital Humanism should also be aware of the critical theory of the Frankfurt School of philosophy. Its prominent members Adorno and Horkheimer provide a critical analysis of the process of empowerment from rationality, and the resulting de-mystification would in principle apply to any technological process that aims to increase the power of the individual. This certainly applies to most digital technologies. But already Habermas, a later member of the Frankfurt School, has pointed out that throwing out rationalism would also mean discarding its many important contributions to law, democracy, and science—and thus also to technology. One can even draw an interesting reinforcing link of Digital Humanism to the dialectic of the Enlightenment: individual decisions are an important source of digital innovation, but this source can also lead to a dangerous gain in power to manipulate masses collectively. In addition, digital humanists also warn about the power of the *knowing caste* (as indicated by Horkheimer and Adorno) or, as a digital humanist might say, the power of *platforms*. And where abstraction was identified as a tool to manipulate, and formulae as tools to create predictability, digital humanists are now wary of digital tools and big data abstractions with very similar concerns. Machine-based abstraction has become a prerequisite for dealing with the complexities of our world. The alignment of such abstractions in Information Technology with human values and with the complexity of our natural environment are core objectives of Digital Humanism, and they remain a task ahead.

Digital Humanism is young, so, understandably, there are different definitions, understandings, and perspectives, and it has different historical roots. In some sense, we are still in the process of theory building, with respect to understanding the interplay between man and modern technology as well as to framing possible approaches to alternative designs. With this book, we take a step into this direction, where we aim to be inclusive and integrative. We invited renowned international colleagues with varying institutional and disciplinary backgrounds to contribute in an open way with their thoughts and ideas. In the end, we received 46 contributions from 60 colleagues, providing their views on the present and future of the digital world. We think that this approach (including also a peer review phase) worked well; the free format approach and the shortness of contributions make for an accessible variety of perspectives. We offer you here the result of this endeavor.

Although no grouping is perfect, we tried to assign the contributions to 11 parts. We start with the *Vienna Manifesto on Digital Humanism*, as it is the basis of our joint undertaking and lays down core principles.

- The first part, *Artificial Intelligence, Humans, and Control*, examines the tension between technology-driven and human-driven decision-making; it looks at the difference between humans and machines and asks whether we are losing control.
- *Participation and Democracy* examines the interplay between digital technologies and democratic practices and takes into consideration diversity issues, also in a geographical context.
- *Ethics and Philosophy of Technology* studies the extent to which digital technologies change our ethical and epistemological perspectives, and also the other way around: what role ethical considerations should play in technology development.
- *Information Technology and the Arts* addresses how the notion of creativity is changed by technology. It connects Digital Humanism with artistic practices as

well as our cultural heritage while highlighting also the importance of culture and art for digital innovation. Science fiction, for example, is a driver of digital innovation.

- *Data, Algorithm, and Fairness* looks at the potential that digital technologies have to both reinforce and ameliorate unfair treatment of groups of humans. It deals with complex questions that may arise from an overly strong focus on the individual rather than a societal perspective. It considers the attention economy and effects that arise from the characteristics of internet search.
- *Platform Power* examines the economic and societal role of today's mega-platforms, such as Google, Facebook, and Twitter, looking at their dynamics, on the important role they played in the pandemic, and their impact in specific industries and their business models.
- *Education and Skills of the Future* considers how the future of work will affect education, the impact of technology on the skills needed in future, and what and how we should teach our young.
- *Digital Geopolitics and Sovereignty* looks at the contradiction of the inherent global dimension of the digital world and the limits of national governance structures. What is the future of sovereignty in digital times?
- *Systems and Society* addresses societal issues such as the future work, how to deal with changes imposed by the digital world, how to frame technological design, and how to formulate corresponding political answers.
- *Learning From Crisis* addresses the role of technology in the human reaction to the global pandemic of 2020–2021, and it draws important lessons for a probable next (global) crisis.
- *Realizing Digital Humanism* reflects on possible next steps and on the level of research, writing on a more general societal and political level. As one contribution states, it seems easy to describe the problem, but it is hard to solve it.

Digital Humanism is a fundamental concept; it is about our future as humans and as society, not only in the digital world. As such, it is not only an academic undertaking, it is also a political issue. We need to engage with society, having a mixed audience, from academics to political decision-makers, from industry and institutions to civil society and non-governmental organizations, and it is not only about science, research, and innovation. Equally important are education, communication, and influencing the public for democratic participation. We hope that this collection of essays provides an essential contribution to this important endeavor.

We want to thank our colleagues for their contributions, also for responding on time (at least most of the time) to our usually "urgent" requests. It was a pleasure to work with you—thank you. We also thank our donors who made this volume possible. We follow an open access strategy, with the content being accessible both via our website as well as being published by Springer. Donors are the City of Vienna (Kulturabteilung), WWTF (Vienna Science and Technology Fund), Austrian Ministry of European and International Affairs, iCyPhy (the Industrial Cyber-Physical Research Center at UC Berkeley), and the Database and Artificial Intelligence Group at TU Wien. Finally, we want to thank Mete Sertkan and

Stephanie Wogowitsch, from the e-commerce group of TU Wien. Without their support and commitment, this undertaking would not have been possible.

The work here is about the need to interfere with the process of digitalization, to change this process. But who will be the agent of change? Our hope is that this book will motivate you, our readers, to contribute and to participate in our journey into the future. In the end, it is up to us, the citizens of the world.

Vienna, Austria   Hannes Werthner
Vienna, Austria   Erich Prem
Berkeley, CA, USA   Edward A. Lee
Milano, Italy   Carlo Ghezzi

Vienna Manifesto on Digital Humanism

VIENNA, MAY 2019

"*The system is failing*"—stated by the founder of the Web, Tim Berners-Lee—emphasizes that while digitalization opens unprecedented opportunities, it also raises serious concerns: the monopolization of the Web, the rise of extremist opinions and behavior orchestrated by social media, the formation of filter bubbles and echo chambers as islands of disjoint truths, the loss of privacy, and the spread of digital surveillance. Digital technologies are disrupting societies and questioning our understanding of what it means to be human. The stakes are high and the challenge of building a just and democratic society with humans at the center of technological progress needs to be addressed with determination as well as scientific ingenuity. Technological innovation demands social innovation, and social innovation requires broad societal engagement.

This manifesto is a call to deliberate and to act on current and future technological development. We encourage our academic communities, as well as industrial leaders, politicians, policy makers, and professional societies all around the globe, to actively participate in policy formation. Our demands are the result of an emerging process that unites scientists and practitioners across fields and topics, brought together by concerns and hopes for the future. We are aware of our joint responsibility for the current situation and the future—both as professionals and citizens.

Today, we experience the co-evolution of technology and humankind. The flood of data, algorithms, and computational power is disrupting the very fabric of society by changing human interactions, societal institutions, economies, and political structures. Science and the humanities are not exempt. This disruption simultaneously creates and threatens jobs, produces and destroys wealth, and improves and damages our ecology. It shifts power structures, thereby blurring the human and the machine.

The quest is for enlightenment and humanism. The capability to automate human cognitive activities is a revolutionary aspect of computer science/informatics. For many tasks, machines surpass already what humans can accomplish in speed, precision, and even analytic deduction. The time is right to bring together humanistic ideals with critical thoughts about technological progress. We therefore link this manifesto to the intellectual tradition of humanism and similar movements striving for an enlightened humanity.

Like all technologies, digital technologies do not emerge from nowhere. They are shaped by implicit and explicit choices and thus incorporate a set of values, norms, economic interests, and assumptions about how the world around us is or should be. Many of these choices remain hidden in software programs implementing algorithms that remain invisible. In line with the renowned Vienna Circle and its contributions to modern thinking, we want to espouse critical rational reasoning and the interdisciplinarity needed to shape the future.

We must shape technologies in accordance with human values and needs, instead of allowing technologies to shape humans. Our task is not only to rein in the downsides of information and communication technologies, but to encourage human-centered innovation. We call for a Digital Humanism that describes, analyzes, and, most importantly, influences the complex interplay of technology and humankind, for a better society and life, fully respecting universal human rights.

In conclusion, *we proclaim the following core principles:*

- *Digital technologies should be designed to promote democracy and inclusion.* This will require special efforts to overcome current inequalities and to use the emancipatory potential of digital technologies to make our societies more inclusive.
- *Privacy and freedom of speech are essential values for democracy and should be at the center of our activities.* Therefore, artifacts such as social media or online platforms need to be altered to better safeguard the free expression of opinion, the dissemination of information, and the protection of privacy.
- *Effective regulations, rules and laws, based on a broad public discourse, must be established.* They should ensure prediction accuracy, fairness and equality, accountability, and transparency of software programs and algorithms.
- *Regulators need to intervene with tech monopolies.* It is necessary to restore market competitiveness as tech monopolies concentrate market power and stifle innovation. Governments should not leave all decisions to markets.
- *Decisions with consequences that have the potential to affect individual or collective human rights must continue to be made by humans.* Decision makers must be responsible and accountable for their decisions. Automated decision making systems should only support human decision making, not replace it.
- *Scientific approaches crossing different disciplines are a prerequisite for tackling the challenges ahead.* Technological disciplines such as computer science/informatics must collaborate with social sciences, humanities, and other sciences, breaking disciplinary silos.
- *Universities are the place where new knowledge is produced and critical thought is cultivated.* Hence, they have a special responsibility and have to be aware of that.

- *Academic and industrial researchers must engage openly with wider society and reflect upon their approaches.* This needs to be embedded in the practice of producing new knowledge and technologies, while at the same time defending the freedom of thought and science.
- *Practitioners everywhere ought to acknowledge their shared responsibility for the impact of information technologies.* They need to understand that no technology is neutral and be sensitized to see both potential benefits and possible downsides.
- *A vision is needed for new educational curricula, combining knowledge from the humanities, the social sciences, and engineering studies.* In the age of automated decision making and AI, creativity and attention to human aspects are crucial to the education of future engineers and technologists.
- *Education on computer science/informatics and its societal impact must start as early as possible.* Students should learn to combine information-technology skills with awareness of the ethical and societal issues at stake.

We are at a crossroads to the future; we must go into action and take the right direction!

SIGN AND SUPPORT THE MANIFESTO:

Authors

Hannes Werthner (TU Wien, Austria), Edward A. Lee (UC Berkeley, USA), Hans Akkermans (Free University Amsterdam, Netherlands), Moshe Vardi (Rice University, USA), Carlo Ghezzi (Politecnico di Milano, Italy), Nadia Magnenat-Thalmann (University of Geneva, Switzerland), Helga Nowotny (Chair of the ERA Council Forum Austria, Former President of the ERC, Austria), Lynda Hardman (CWI, Centrum Wiskunde & Informatica, Netherlands), Oliviero Stock (Fondazione Bruno Kessler, Italy), James Larus (EPFL, Switzerland), Marco Aiello (University of Stuttgart, Germany), Enrico Nardelli (Università degli Studi di Roma "Tor Vergata", Italy), Michael Stampfer (WWTF, Vienna Science and Technology Fund, Austria), Christopher Frauenberger (TU Wien, Austria), Magdalena Ortiz

(TU Wien, Austria), Peter Reichl (University of Vienna, Austria), Viola Schiaffonati (Politecnico di Milano, Italy), Christos Tsigkanos (TU Wien, Austria), William Aspray (University of Colorado Boulder, USA), Mirjam E. de Bruijn (Leiden University, Netherlands), Michael Strassnig (WWTF, Vienna Science and Technology Fund, Austria), Julia Neidhardt (TU Wien, Austria), Nikolaus Forgo (University of Vienna, Austria), Manfred Hauswirth (TU Berlin, Germany), Geoffrey G. Parker (Dartmouth College, USA), Mete Sertkan (TU Wien, Austria), Allison Stanger (Middlebury College & Santa Fe Institute, USA), Peter Knees (TU Wien, Austria), Guglielmo Tamburrini (University of Naples, Italy), Hilda Tellioglu (TU Wien, Austria), Francesco Ricci (Free University of Bozen-Bolzano, Italy), Irina Nalis-Neuner (University of Vienna, Austria)

Contents

Part I
Artificial Intelligence, Humans, and Control

Are We Losing Control?

Edward A. Lee

Abstract This chapter challenges the predominant assumption that humans shape technology using top-down, intelligent design, suggesting that technology should instead be viewed as the result of a Darwinian evolutionary process where humans are the agents of mutation. Consequently, we humans have much less control than we think over the outcomes of technology development.

Rapid change breeds fear. With its spectacular rise from the ashes in the last decade, we fear that AI may replace most white collar jobs (Ford 2015); that it will learn to iteratively improve itself into a superintelligence that leaves humans in the dust (Barrat 2013; Bostrom 2014; Tegmark 2017); that it will fragment information so that humans divide into islands of disjoint sets of truths (Lee 2020); that it will supplant human decision-making in health care, finance, and politics (Kelly 2016); that it will cement authoritarian powers, tracking every move of their citizens and shaping their thoughts (Lee 2018); and that the surveillance capitalists' monopolies, which depend on AI, will destroy small business and swamp entrepreneurship (Zuboff 2019).

Surely, today, we still retain a modicum of control. At the very least, we can still pull the plug. Or can we? The technology underlying these risks is made by humans, so why can't we control the outcomes? We have the power to design and to regulate, don't we? So why are we trying so desperately to catch up with yesterday's disasters while today's just fester? The very technology that threatens us is also the reason we are successfully feeding most of the 7.8 billion humans on this meager planet and have lifted billions out of poverty in the last decades. Giving us pause, however, Albert Einstein famously said, "We cannot solve our problems with the same thinking we used when we created them."

Knowledge is at the root of technology, information is at the root of knowledge, and today's technology makes information vastly more accessible than it has ever

E. A. Lee (✉)
University of California, Berkeley, CA, USA
e-mail: eal@berkeley.edu

been. Shouldn't this help us solve our problems? The explosion of AI feeds the tsunami, turning every image, every text, and every sound into yet more information, flooding our feeble human brains. We can't absorb the flood without curation, and curation of information is increasingly being done by AIs. Every subset of the truth is only a partial truth, and curated information includes, necessarily, a subset. Since our brains can only absorb a tiny subset of the flood, everything we take in is at best a partial truth. The AIs, in contrast, seem to have little difficulty with the flood. To them, it is the food that strengthens, perhaps leading to that feared runaway feedback loop of superintelligence that sidelines humans into irrelevance.

The question I address here is, "Are we going to lose control?" You may find my answer disturbing.

First, in posing this question, what do we mean by "we"? Do we mean "humanity," all 7.8 billion of us? The idea of 7.8 billion people collectively controlling anything is patently absurd, so that must not be what we mean. Do we mean the engineers of Silicon Valley? The investors on Wall Street? The politicians who feed off the partial truths and overt lies?

Second, what do we mean by "control"? Is it like steering a car on a network of roads, or is it more like steering a car while the map emerges and morphs into unexpected dead ends, underpasses, and loops? If we are steering technology, then every turn we take changes the terrain that we have to steer over in unexpected ways.

I am an engineer. In my own small way, I contribute to the problem by writing software, some of which has small influences on our ecosystem. For most of my 40 years doing this, I harbored a "creationist" illusion that the things I designed were my own personal progeny, the pure result of my deliberate decisions, my own creative output. I have since realized that this is a bit like thinking that the bag of groceries that I just brought back from the supermarket is my own personal accomplishment. It ignores centuries of development in the technology of the car that got me there and back, agriculture that delivered the incredible variety of fresh food to the store, the economic system that makes all of this affordable, and many other parts of the socio-cultural backdrop against which my meager accomplishment pales.

In my recent book (Lee 2020), I coin the term "digital creationism" for the idea that technology is the result of top-down intelligent design. This principle assumes that every technology is the outcome of a deliberate process, where every aspect of a design is the result of an intentional, human decision. I now know, 40 years later, that this is not how it happens. Software engineers are more the agents of mutation in a Darwinian evolutionary process. The outcome of their efforts is shaped more by the computers, networks, software tools, libraries, programming languages, and other programs they use than by their deliberate decisions. And the success and further development of their product is determined as much or more by the cultural milieu into which they launch their "creation" than by their design decisions.

The French philosopher known as Alain (whose real name was Émile-Auguste Chartier), wrote, about fishing boats in Brittany:

> Every boat is copied from another boat. ... Let's reason as follows in the manner of Darwin. It is clear that a very badly made boat will end up at the bottom after one or two voyages and thus never be copied. ... One could then say, with complete rigor, that it is the sea herself who fashions the boats, choosing those which function and destroying the others. (Rogers and Ehrlich 2008)

Boat designers are agents of mutation, and sometimes their mutations result in a badly made boat. From this perspective, perhaps Facebook has been fashioned more by teenagers than by software engineers.

More deeply, digital technology *co*evolves with humans. Facebook changes its users, who then change Facebook. For software engineers, the tools we use, themselves earlier outcomes of software engineering, shape our thinking. Think about how IDEs[1] (such as Eclipse or Visual Studio Code), message boards (such as Stack Overflow), libraries (such as the Standard Template Library), programming languages (e.g., Scala, Rust, and JavaScript), and Internet search (such as Google or Bing) affect the outcome of our software. These tools have more effect on the outcome than all of our deliberate decisions.

Today, the fear and hype around AI taking over the world and social media taking down democracy has fueled a clamor for more regulation. But if I am right about coevolution, we may be going about the project of regulating technology all wrong. Why have privacy laws, with all their good intentions, done little to protect our privacy? They have only overwhelmed us with small-print legalese and annoying pop-ups giving us a choice between "accept our inscrutable terms" and "go away." Do we expect new regulations trying to mitigate fake news or to prevent insurrections from being instigated by social media to be any more effective?

Under the principle of digital creationism, bad outcomes are the result of unethical actions by individuals, for example by blindly following the profit motive with no concern for societal effects. Under the principle of coevolution, bad outcomes are the result of the "procreative prowess" (Dennett 2017) of the technology itself. Technologies that succeed are those that more effectively propagate. The individuals we credit with (or blame for) creating those technologies certainly play a role, but so do the users of the technologies and their whole cultural context. Under this perspective, Facebook users bear some of the blame, along with Mark Zuckerberg, for distorted elections. They even bear some of the blame for the design of Facebook software that enables distorted elections. If they were happy to pay for social networking, for example, an entirely different software design may have emerged.

Under digital creationism, the purpose of regulation is to constrain the individuals who develop and market technology. In contrast, under coevolution, constraints can be about the *use* of technology, not just its design and the business of selling it. The purpose of regulation becomes to nudge the process of both technology and cultural evolution through incentives and penalties. Nudging is probably the best we can

---

[1] Integrated development environments (IDEs) are computer programs that assist programmers by parsing their text as they type, coloring text by function, identifying errors and potential flaws in code style, suggesting insertions, and transforming code through refactoring.

hope for. Evolutionary processes do not yield easily to control because the territory over which we have to navigate keeps changing.

Perhaps privacy laws have been ineffective because they are based on digital creationism as a principle. These laws assume that changing the behavior of corporations and engineers will be sufficient to achieve privacy goals (whatever those are for you). A coevolutionary perspective understands that users of technology will choose to give up privacy even if they are explicitly told that their information will be abused. We are repeatedly told exactly that in the fine print of all those privacy policies we don't read, and, nevertheless, our kids get sucked into a media milieu where their identity gets defined by a distinctly non-private online persona.

If technology is defining culture while culture is defining technology, we have a feedback loop, and intervention at any point in the feedback loop can change the outcomes. Hence, it may be just as effective to pass laws that focus on educating the public, for example, as it is to pass laws that regulate the technology producers. Perhaps if more people understood that Pokémon GO is a behavior-modification engine, they would better understand Niantic's privacy policy and its claim that their product, Pokémon GO, has no advertising. Establishments pay Niantic for placement of a Pokémon nearby to entice people to visit them (Zuboff 2019). Perhaps a strengthening of libel laws, laws against hate speech, and other refinements to first-amendment rights should also be part of the remedy.

I believe that, as a society, we can do better than we are currently doing. The risk of an Orwellian state (or perhaps worse, a corporate Big Brother) is very real. It has happened already in China. We will not do better, however, until we abandon digital creationism as a principle. Outlawing specific technology developments will not be effective, and breaking up monopolies could actually make the problem worse by accelerating mutations. For example, we may try to outlaw autonomous decision-making in weapons systems and banking, but as we see from election distortions and Pokémon GO, the AIs are very effective at influencing human decision-making, so putting a human in the loop does not necessarily help. How can a human who is, effectively, controlled by a machine somehow mitigate the evilness of autonomous weapons?

When I talk about educating the public, many people immediately gravitate to a perceived silver bullet, that we should teach ethics to engineers. But I have to ask, if we assume that all technologists behave ethically (whatever your meaning of that word), can we conclude that bad outcomes will not occur? This strikes me as naïve. Coevolutionary processes are much too complex.

This chapter is my small contribution to the digital humanism initiative, a movement that seeks a more human-centric approach to technology. This initiative makes it imperative for intellectuals of all disciplines to step up and take seriously humanity's dance with technology. That our limited efforts to rein in the detrimental effects of digital technology have been mostly ineffective underscores our weak understanding of the problem. We need humanists with a deeper understanding of technology, technologists with a deeper understanding of the humanities, and policy makers drawn from both camps. We are quite far from that goal today.

Returning to the original question, are we losing control? The answer is "no." We never had control, and we can't lose what we don't have. This does not mean we should give up, however. We *can* nudge the process, and even a supertanker can be redirected by gentle nudging.

References

Barrat, J. (2013). *Our Final Invention: Artificial Intelligence and the End of the Human Era*, St. Martin's Press.

Bostrom, N. (2014). *Superintelligence: Paths, Dangers, Strategies.* Oxford, UK, Oxford University Press.

Dennett, D. C. (2017). *From Bacteria to Bach and Back: The Evolution of Minds*, W. W. Norton and Company.

Ford, M. (2015). *Rise of the Robots -- Technology and the Threat of a Jobless Future.* New York, Basic Books.

Kelly, K. (2016). *The Inevitable: Understanding the 12 Technological Forces That Will Shape Our Future.* New York, Penguin Books.

Lee, E. A. (2020). *The Coevolution: The Entwined Futures of Humans and Machines.* Cambridge, MA, MIT Press.

Lee, K.-F. (2018). *Super-Powers: China, Silicon Valley, and the New World Order.* New York, Houghton Mifflin Harcourt Publishing Company.

Rogers, D. S. and P. R. Ehrlich (2008). "Natural Selection and Cultural Rates of Change." *Proceedings of the National Academy of Sciences of the United States of America* 105(9): 3416-3420.

Tegmark, M. (2017). *Life 3.0: Being Human in the Age of Artificial Intelligence.* New York, Alfred A. Knopf.

Zuboff, S. (2019). *The Age of Surveillance Capitalism: The Fight for a Human Future at the New Frontier of Power*, PublicAffairs , Hachette Book Group.

Social Robots: Their History and What They Can Do for Us

Nadia Magnenat Thalmann

Abstract From antiquity to today, some scientists, writers, and artists are passionate about representing humans not only as beautiful statues but as automatons that can perform actions. Already in ancient Greece, we can find some examples of automatons that replaced servants. In this chapter, we go through the development of automatons until the social robots of today. We describe two examples of social robots, EVA and Nadine, that we have been working with. We present two case studies, one in an insurance company and the other one in an elderly home. We also mention the limits of the use of social robots, their dangers, and the importance to control their actions through ethical committees.

Talos, the giant automaton from Greece

N. M. Thalmann (✉)
MIRALab, University of Geneva, Geneva, Switzerland
e-mail: Nadia.thalmann@unige.ch

1 History of Human Robotics

Fascination for automation is neither new nor modern. Living side by side with automated, mechanical beings which look like us and can do fantastic things or protect us is one of mankind's oldest dreams. In 400 BC, Greek mythology already featured the tale of Talos, a 10-m giant automaton made of bronze. He was brought to life by the gods to ward off pirates and invaders and to keep watch to protect the island. Sometime later, in 250 BC, a human-like automatic device in the form of a maid, holding a jug of wine in her right hand was made. When one placed a cup in the palm of the automaton's left hand, it automatically poured wine initially and then poured water into the cup, mixing it when desired (Fig. 1).

This automatic maid can be seen in the Museum of ancient Greek Technology in Katalo in Greece.

Leonardo Da Vinci, in the late fifteenth century, did technical drawings that were found recently. These drawings have allowed the creation of a mechanical humanoid automaton, a robotic knight which is capable of independent motion. It could stand, sit, raise its visor, and maneuver its arms independently (Fig. 2). It was fully done according to Leonardo Da Vinci's drawings and is fully working.

Later, a marvel of automatons has been produced and shown in 1775 in London. Spectators could see three animated human figures made by Pierre and Henri Louis Droz, from Switzerland. These automatons were autonomously able to write, to

Fig. 1 Automatic maid, done by Philon

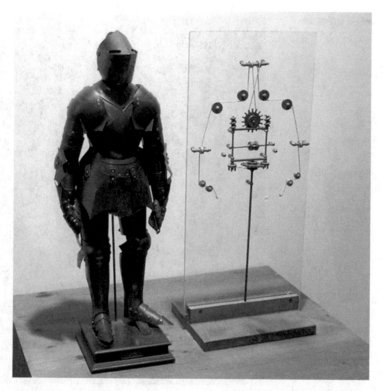

Fig. 2 Robotic knight

draw, or to play a keyboard instrument. This exhibition at Covent Garden generated a lot of questions among famous writers. Some imaginary plays as, for example, "Les Contes d'Hoffmann" were produced. Hoffmann is the author of one of the masterpieces "The Sandman." In this piece, he tells the story of a young man who fell in love with the automaton Olympia. It has been followed by Jacques Offenbach's opera, "Les Contes d'Hoffmann." These imaginary masterpieces, even if they demonstrate no new technical real automaton, show how someone can be duped by the beauty and actions of an automaton (Fig. 3).

The development of automations has created other philosophical interests. In the eighteenth century, French philosopher Denis Diderot reflected on the possibility of modelling intelligence. In his *Pensées Philosophiques*, he wrote "If they find a parrot who could answer everything, I will claim it to be an intelligent being without hesitation." As early as the eighteenth century, there was already a lot of discussion about what intelligence is and whether it could be modelled to be integrated within machines.

Fig. 3 Swiss automaton

In 1950, Alan Turing developed the Turing Test to test a machine's ability to exhibit intelligent behavior equivalent to, or indistinguishable from, that of a human.[1] However, it was soon felt that the Turing Test was not sufficient. As the notions of social and emotional intelligence theory grew in the 1980s and 1990s, human intelligence was understood not only as an ability to answer logical questions based on logical reasoning, but rather as an ability to take account of one's environment, the real world, social interactions, and emotions, and this is something the Turing Test could not measure. This paved the way for new technical developments towards intelligent social robots.

Alongside the evolution in social sciences, computer science, too, developed over time. Sixty years ago, computers were very large and were very limited. Today, they are much smaller, much faster, and much more powerful and offer incredible possibilities of interfacing with people through sensors and actuators. We have now many software and hardware tools which can capture, understand, and analyze a lot of signals and meanings. We can capture speech, sounds, gestures, shapes, etc. Through the emergence of big data and machine learning, we can analyze new data, find correspondence, and predict future patterns.

---

[1] https://en.wikipedia.org/wiki/Turing_test

Fig. 4 EVA interacting
with a user

2 The Challenges of Becoming Social

A social robot is defined as "an autonomous entity that interacts and communicates with humans or other autonomous physical agents by following social behaviours and rules attached to its role."[2] Social robots interact with people in social contexts. Therefore, they must understand the social context, i.e., understand users' behaviors and emotions, and respond with the appropriate gestures, facial expressions, and gaze. The challenge is to provide algorithms to sense, analyze situations and intentions, and make appropriate decisions.

Our lab's first social robot project in Switzerland was EVA (2008–2012), a robotic tutor (Fig. 4). The overall goal of this project was to develop a long-term social interaction framework with a human-like robot or virtual human, modelling emotions, episodic memory, and expressive behavior.

EVA can interact with users by recognizing and remembering their names and can understand users' emotional states through facial expression recognition and speech. Based on user input and its personality model, EVA produces appropriate emotional responses and keeps information on the long-term interpersonal relationships (Kasap and Magnenat Thalmann 2012). EVA played a role as an actor with real actors in the Roten Fabrik Theater at Zurich.

Taking this one step further, in 2013, we started in NTU in Singapore working on Nadine, a socially intelligent robot with a strong human-likeness, a natural-looking skin and hair, and realistic hands and body. One of our main motivation over time is not only to produce a technical innovation but also to contribute to art. Nadine has been recognized as a beautiful living sculpture, a piece of expressive art with a very human-computer natural interface.[3] We consider this kind of realistic robots as the next step in the making of living sculptures. Nadine is equipped with a 3D camera and a microphone. It can recognize people, gestures, and clues of social situations of emotion and behavior as well as speech, which enables it to understand social situations. Nadine has an emotional module, a memory, and a social attention

---

[2] https://en.wikipedia.org/wiki/Social_robot

[3] https://en.wikipedia.org/wiki/Nadine_Social_Robot

Fig. 5 Nadine and Nadia

model as well as a chatbot, which means that depending on the input, it will generate the appropriate emotional response and remember interactions. Nadine's controllers control lip synchronization and its gaze and enable Nadine to adapt its facial expressions to basic emotions (Fig. 5). It has a personality, mood, and emotion model, and depending on its relationship with a user, its response will vary and adapt to each situation.

3 Case Studies in an Insurance Company and in an Elderly Home

We have conducted a study with Nadine humanoid social robot as a customer service agent in an insurance company in Singapore alongside human employees. Nadine was accessible to the public, and the customer could ask any question relative to the insurance. As a service customer agent, Nadine was able to answer customer queries and maintain a courteous, professional behavior during interaction. Our objective was to study if a robot could handle real social situations, if customers were willing to interact with human-like robots, and to study the usefulness of a humanoid robot

Fig. 6 Nadine at AIA Insurance

in an insurance environment. For customer queries, the company had provided several FAQs from their customer interactions. A separate chatbot was trained based on these FAQs and integrated into Nadine, which allowed her to handle customer queries.

The appearance of the robot played a vital role in the way customers perceived the robot. The company trained at the same time another robot named Pepper. The study showed clearly that Nadine was preferred as it had a human appearance and was more trustable as a customer agent than the robot Pepper that had a more robotic appearance. Pepper was considered as a fancy robot to play with but not for answering and informing people how to get an insurance policy. Nadine was able to provide a better customer interaction experience (Fig. 6).

With the combined effect of increasing life spans, reduced mortality rates, and declining birth rates, the world is growing older. Therefore, the need for social or assistive robots will become more pressing than ever before. A report published by the World Economic Forum states that "Current projections already estimate that by 2050, Japan will have 72 dependent people over 65 for every 100 workers. For Spain, Italy, and Germany, these numbers range between 60% and 70%." In such contexts, social robots will become a necessary and vital ally in managing our ageing society in a sustainable way.

We have just completed a 6-month study in Bright Hill Evergreen Home, an elderly home in Singapore where Nadine has played bingo with the elderly and interacted with them upon their demand. The data analysis is very positive. Patients largely prefer having a social human-like robot to play bingo with them instead of a therapist who has little time. Nadine could repeat the numbers, show results on several screens, and take the time to congratulate people. Patients also highly appreciated the natural human-like beauty of Nadine.[4]

4   Ethical Issues of Social Robots

Most researchers are developing robots for a better humanity, and the examples cited above fall into this category. Others fabricate killer robots, dominant robots, and bio robots to use them against humans for profits or for hateful purposes. Fortunately, researchers from diverse areas are beginning to formulate together ethical questions about the use of robotic technology and its implementation in societies. A very important report on the ethics of robotics has been published by the World Commission of the Ethics of Scientific Knowledge and Technology in 2017.[5] Important Ethical committees are taken place as IEEE and ACM[6] and other professional organizations. More importantly, in research, any project on social robots interacting with humans needs an approval from institutional review boards or ethical committees.

I would like to conclude on the following citation by Aristotle who speculated in his *Politics* (ca. 322 BC, book 1, part 4) that automata could someday bring human equality:

> If, in like manner, the shuttle would weave and the plectrum touch the lyre without a hand to guide them, chief workmen would not want servants, nor masters' slaves.

Reference

Kasap, Z. and Magnenat Thalmann, N. (2012). Building Long-term Relationships with Virtual and Robotic Characters: The Role of Remembering, The Visual Computer, vol. 28, no. 1, pp. 87-97.

---

[4] Do Elderly enjoy playing Bingo with a humanoid robot? Submitted at Ro-Man 2021 Conference, https://www.dropbox.com/s/fj7qdqm4a01ezhz/NT_Humanoid%20man%20vs%20robots%20Final%20reviewed_May%203.docx?dl=0

[5] https://unesdoc.unesco.org/ark:/48223/pf0000253952 (report on the ethics of robotics)

[6] https://dl.acm.org/doi/10.1145/1167867.1164071 (Ethical acts in robotics report)

Artificial Intelligence and the Problem of Control

Stuart Russell

Abstract A long tradition in philosophy and economics equates intelligence with the ability to act rationally—that is, to choose actions that can be expected to achieve one's objectives. This framework is so pervasive within AI that it would be reasonable to call it the standard model. A great deal of progress on reasoning, planning, and decision-making, as well as perception and learning, has occurred within the standard model. Unfortunately, the standard model is unworkable as a foundation for further progress because it is seldom possible to specify objectives completely and correctly in the real world. The chapter proposes a new model for AI development in which the machine's uncertainty about the true objective leads to qualitatively new modes of behavior that are more robust, controllable, and deferential to humans.

1 The Standard Model

The central technical concept in AI is that of an *agent*—an entity that perceives and acts (Russell and Norvig 2020).[1] Cognitive faculties such as reasoning, planning, and learning are in the service of acting. The concept can be applied to humans, robots, software entities, corporations, nations, or thermostats. AI is concerned principally with designing the internals of the agent: mapping from a stream of raw perceptual data to a stream of actions. Designs for AI systems vary enormously depending on the nature of the environment in which the system will operate, the nature of the perceptual and motor connections between agent and environment, and the requirements of the task.

AI seeks agent designs that exhibit "intelligence," but what does this mean? Aristotle (*Ethics*) gave one answer: "We deliberate not about ends, but about means. ... [We] assume the end and consider how and by what means it is attained,

---

[1] The word "agent" in AI carries no connotation of acting on behalf of another.

S. Russell (✉)
University of California, Berkeley, CA, USA
e-mail: russell@berkeley.edu

© The Author(s) 2022
H. Werthner et al. (eds.), *Perspectives on Digital Humanism*,
https://doi.org/10.1007/978-3-030-86144-5_3

and if it seems easily and best produced thereby." That is, an intelligent or *rational* action is one that can be expected to achieve one's objectives. This line of thinking has persisted to the present day. Arnauld (1662) broadened Aristotle's theory to include uncertainty in a quantitative way, proposing that we should act to maximize the *expected value* of the outcome. Daniel Bernoulli (1738) refined the notion of value, moving it from an external quantity (typically money) to an internal quantity that he called *utility*. De Montmort (1713) noted that in games (decision situations involving two or more agents) a rational agent might have to act randomly to avoid being second-guessed. Von Neumann and Morgenstern (1944) tied all these ideas together into an axiomatic framework that underlies much of modern economic theory.

As AI emerged in the 1940s and 1950s, it needed some notion of intelligence on which to build the foundations of the field. Although some early research was aimed more at emulating human cognition, the notion that won out was rationality: a machine is intelligent to the extent that its actions can be expected to achieve its objectives. In the standard model, we aim to build machines of this kind; we define the objectives; and the machine does the rest. There are several different ways in which the standard model can be instantiated. For example, a problem-solving system for a deterministic environment is given a cost function and a goal criterion and finds the least-cost action sequence that leads to a goal state; a reinforcement learning system for a stochastic environment is given a reward function and a discount factor and learns a policy that maximizes the expected discounted sum of rewards.

This general approach is not unique to AI. Control theorists minimize cost functions; operations researchers maximize rewards; statisticians minimize an expected loss function; and economists, of course, maximize the utility of individuals, the welfare of groups, or the profit of corporations.

In short, the standard model of AI (and related disciplines) is a pillar of twentieth-century technology.

2 Difficulties of the Standard Model

Unfortunately, the standard model is unworkable as a foundation for further progress. Once AI systems move out of the laboratory (or artificially defined environments such as the simulated chessboard) and into the real world, there is very little chance that we can specify our objectives completely and correctly in such a way that the pursuit of those objectives by more capable machines is guaranteed to result in beneficial outcomes for humans. Indeed, we may lose control altogether, as noted by Turing (1951): "*It seems probable that once the machine thinking method had started, it would not take long to outstrip our feeble powers. … At some stage therefore we should have to expect the machines to take control.*" We can expect a sufficiently capable machine pursuing a fixed objective to take preemptive steps to ensure that the stated objective is achieved, including acquiring physical and computational resource and defending against any possible attempt to interfere with goal achievement.

The Vienna Manifesto on Digital Humanism includes the following principle: "We must shape technologies in accordance with human values and needs, instead of allowing technologies to shape humans." Perhaps the clearest example demonstrating the need for this principle is given by machine learning algorithms performing content selection on social media platforms. Such algorithms typically pursue the objective of maximizing clickthrough or a related metric. Rather than simply adjusting their recommendations to suit human preferences, these algorithms will, in pursuit of their long-term objective, learn to manipulate humans to make them more predictable in their clicking behavior (Groth et al. 2019).[2] This effect may be contributing to growing polarization and extremism in many countries.

The mistake in the standard model comes from transferring a perfectly reasonable definition of intelligence from humans to machines. The definition is reasonable for humans because we are entitled to pursue our own objectives. (Indeed, whose would we pursue if not our own?) Machines, on the other hand, are not entitled to pursue their own objectives. A more sensible definition of AI would have machines pursuing *our* objectives. In the unlikely event that we can specify the objectives completely and correctly and insert them into the machine, we can recover the standard model as a special case. If not, then the machine will necessarily be uncertain as to our objectives, while being obliged to pursue them on our behalf. This uncertainty—with the coupling between machines and humans that it entails— turns out to be crucial to building AI systems of arbitrary intelligence that are *provably beneficial* to humans. In other words, I propose to do more than "shape technologies in accordance with human values and needs." Because we cannot necessarily articulate those values and needs, we must design technologies that will, by their very constitution, respond to human values and needs, whatever they are.

3 A New Model

In *Human Compatible* (Russell 2019), I suggest three principles underlying a new model for creating AI systems:

1. The machine's only objective is to maximize the realization of human preferences.
2. The machine is initially uncertain about what those preferences are.
3. The ultimate source of information about human preferences is human behavior.

---

[2]Providing additional evidence for the significance of the problem of misspecified objectives, Hillis (2019) has drawn the analogy between uncontrollable AI systems and uncontrollable economic actors—such as fossil-fuel corporations maximizing profit at the expense of humanity's future.

As noted in the preceding section, the uncertainty about objectives that the second principle espouses is a relatively unstudied concept in AI—yet it is central to ensuring that we not lose control over increasingly capable AI systems.

In the 1980s, the AI community abandoned the idea that AI systems could have definite knowledge of the state of the world or of the effects of actions, and they embraced uncertainty in these aspects of the problem statement. It is not at all clear why, for the most part, they failed to notice that there must also be uncertainty in the objective. Although some AI problems such as puzzle solving are designed to have well-defined goals, many other problems that were considered at the time, such as recommending medical treatments, have no precise objectives and ought to reflect the fact that the relevant preferences (of patients, relatives, doctors, insurers, hospital systems, taxpayers, etc.) are not known initially in each case. While it is true that unresolvable uncertainty over objectives can be integrated out of any decision problem, leaving an equivalent decision problem with a definite (average) objective, this transformation is invalid when there is the possibility of additional evidence regarding the true objectives. Thus, one may characterize the primary difference between the standard and new models of AI through the flow of preference information from humans to machines at "runtime." This flow comes from evidence provided by human behavior, as the third principle asserts.

This basic idea is made more precise in the framework of assistance games—originally known as cooperative inverse reinforcement learning (CIRL) games in the terminology of Hadfield-Menell et al. (2017a). The simplest case of an assistance game involves two agents, one human and the other a robot. It is a game of partial information, because, while the human (in the basic version) knows the payoff function, the robot does not—even though the robot's job is to maximize it. In a Bayesian formulation, the robot begins with a prior probability distribution over the human payoff function and updates it as the robot and human interact during the game. The basic assistance game model can be elaborated to allow for imperfectly rational humans (Hadfield-Menell et al. 2017b), humans who don't know their own preferences (Chan et al. 2019), multiple human participants (Fickinger et al. 2020), multiple robots, and so on.

Assistance games are connected to inverse reinforcement learning, or IRL (Russell 1998; Ng and Russell 2000), because the robot can learn more about human preferences from the observation of human behavior—a process that is the dual of reinforcement learning, wherein behavior is learned from rewards and punishments. The primary difference is that in the assistance game, unlike the IRL framework, the human's actions are affected by the robot's presence—for example, the human may try to teach the robot about his or her preferences. This two-way process lends the framework an inevitable game-theoretic character that produces, among other phenomena, emergent conventions for communicating preference information.

The overall approach also resembles principal–agent problems in economics, wherein the principal (e.g., an employer) needs to incentivize another agent (e.g., an employee) to behave in ways beneficial to the principal. The key difference here is that unlike a human employee, the robot has no interests of its own. Furthermore, we are building one of the agents in order to benefit the other, so the appropriate solution concepts may differ.

Within the framework of assistance games, a number of basic results can be established that are relevant to Turing's problem of control.

- Under certain assumptions about the support and bias of the robot's prior probability distribution over human rewards, one can show that a robot solving an assistance game has non-negative value to humans (Hadfield-Menell et al. 2017a).
- A robot that is uncertain about the human's preferences has a non-negative incentive to allow itself to be switched off (Hadfield-Menell et al. 2017b). In general, it will defer to human control actions.
- To avoid changing attributes of the world whose value is unknown, the robot will generally engage in "minimally invasive" behavior to benefit the human (Shah et al. 2019). Even when it knows nothing at all about human preferences, it will still take "empowering" actions that expand the set of actions available to the human.

There are too many open research problems in the new model of AI to list them all here. The most directly relevant to moral philosophy and the social sciences is the question of *social aggregation*: how should a machine decide when its actions affect the interests of more than one human being? Issues include the preferences of evil individuals (Harsanyi 1977); relative preferences and positional goods (Veblen 1899; Hirsch 1977); and interpersonal comparison of preferences (Nozick 1974; Sen 1999). Also of great importance is the *plasticity* of human preferences, which brings up both the philosophical problem of how to decide on behalf of a human whose preferences change over time (Pettigrew 2020) and the practical problem of how to ensure that AI systems are not incentivized to change human preferences in order to make them easier to satisfy.

Assuming that the theoretical and algorithmic foundations of the new model for AI can be completed and then instantiated in the form of useful systems such as personal digital assistants or household robots, it will be necessary to create a technical consensus around a set of design templates for provably beneficial AI, so that policy makers have some concrete guidance on what sorts of regulations might make sense. The economic incentives would tend to support the installation of rigorous standards at the early stages of AI development, because failures would be damaging to entire industries, not just to the perpetrator and victim.

The question of *enforcing* policies for beneficial AI is more problematic, given our lack of success in containing malware. In Samuel Butler's *Erewhon* and in Frank Herbert's *Dune*, the solution is to ban all intelligent machines, as a matter of both law and cultural imperative. Perhaps if we find institutional solutions to the malware problem, we will be able to devise some less drastic approach for AI. As the Manifesto underscores, the technology of AI has no value in itself beyond its ability to benefit humanity.

References

Aristotle (n.d.). Nicomachean Ethics, Book III, 3, 1112b.

Arnauld, A. (1662). *La logique, ou l'art de penser*. Paris: Chez Charles Savreux.

Bernoulli, D. (1738). Specimen theoriae novae de mensura sortis. *Proceedings of the St. Petersburg Imperial Academy of Sciences*, 5, 175–92.

Chan, L., Hadfield-Menell, D., Srinivasa, S., & Dragan, A. (2019). The assistive multi-armed bandit. In *Proc. Fourteenth ACM/IEEE International Conference on Human–Robot Interaction*.

De Montmort, P. R. (1713). *Essay d'analyse sur les jeux de hazard*, 2nd ed. Paris: Chez Jacques Quillau.

Fickinger, A., Hadfield-Menell, D., Critch, A., & Russell, S. (2020). Multi-Principal Assistance Games: Definition and Collegial Mechanisms. In *Proc. NeurIPS Workshop on Cooperative AI*.

Groth, O., Nitzberg, M., & Russell, S. (2019, August 15). AI algorithms need FDA-style drug trials. *Wired*.

Hadfield-Menell, D., Dragan, A. D., Abbeel, P., & Russell, S. (2017a). Cooperative inverse reinforcement learning. In *Advances in Neural Information Processing Systems 29*.

Hadfield-Menell, D., Dragan, A. D., Abbeel, P., & Russell, S. (2017b). The off-switch game. In *Proc. Twenty-Sixth International Joint Conference on Artificial Intelligence*.

Harsanyi, J. (1977). Morality and the theory of rational behavior. *Social Research*, 44, 623–656.

Hillis, D. (2019). The first machine intelligences. In John Brockman (ed.), *Possible Minds: Twenty-Five Ways of Looking at AI*. Penguin Press.

Hirsch, F. (1977). *The Social Limits to Growth*. Routledge & Kegan Paul.

Ng, A. Y. & Russell, S. (2000). Algorithms for inverse reinforcement learning. In *Proc. Seventeenth International Conference on Machine Learning*.

Nozick, R. (1974). *Anarchy, State, and Utopia*. Basic Books.

Pettigrew, R. (2020). *Choosing for Changing Selves*. Oxford University Press.

Russell, S. (1998). Learning agents for uncertain environments. In *Proc. Eleventh ACM Conference on Computational Learning Theory*.

Russell, S. (2019). *Human Compatible: AI and the Problem of Control*. London: Penguin.

Russell, S. & Norvig, P. (2020). *Artificial Intelligence: A Modern Approach* (4th edition). Pearson.

Sen, A. (1999). The Possibility of Social Choice. *American Economic Review*, 89, 349–378.

Shah, R., Krasheninnikov, D., Alexander, J., Abbeel, P., & Dragan, A. (2019). The implicit preference information in an initial state. In *Proc. Seventh International Conference on Learning Representations*.

Turing, A. (1951). "Can digital machines think?" Radio broadcast, BBC Third Programme. Typescript available at turingarchive.org.

Veblen, T. (1899). *The Theory of the Leisure Class: An Economic Study of Institutions*. Macmillan.

von Neumann, J. and Morgenstern, O. (1944). Theory of Games and Economic Behavior. Princeton University Press.

The Challenge of Human Dignity in the Era of Autonomous Systems

Paola Inverardi

Abstract Autonomous systems make decisions independently or on behalf of the user. This will happen more and more in the future, with the widespread use of AI technologies in the fabric of the society that impacts on the social, economic, and political sphere. Automating services and processes inevitably impacts on the users' prerogatives and puts at danger their autonomy and privacy. From a societal point of view, it is crucial to understand which is the space of autonomy that a system can exercise without compromising laws and human rights. Following the European Group on Ethics in Science and New Technologies 2018 recommendation, the chapter addresses the problem of preserving the value of human dignity in the context of the digital society, understood as the recognition that a person is worthy of respect in her interaction with autonomous technologies. A person must be able to exercise control on information about herself and on the decisions that autonomous systems make on her behalf.

Nowadays, citizens continuously interact with software systems, e.g., by using a mobile device, in their smart homes, or from on board of a (autonomous) car. This will happen more and more in the future, with the widespread use of artificial intelligence (AI) technologies in the fabric of society that impacts on the social, economic, and political spheres. Effectively described by Floridi's metaphor of the mangrove society (Floridi 2018), the digital world will be increasingly dominated by autonomous systems (AS) that make decisions independently or on behalf of the users. Automating services and processes inevitably impacts on the users' prerogatives and puts at danger their autonomy and privacy.

Besides the known risks represented by, e.g., unauthorized disclosure and mining of personal data or access to restricted resources, which are receiving a huge amount of attention, there is a less evident but more serious risk which attains the core of the fundamental rights of the citizens. Worries about the growing of the data economy

P. Inverardi (✉)
Dipartimento di Scienze ed Ingegneria dell'Informazione e Matematica, Università dell'Aquila, L'Aquila, Italy
e-mail: paola.inverardi@univaq.it

© The Author(s) 2022
H. Werthner et al. (eds.), *Perspectives on Digital Humanism*,
https://doi.org/10.1007/978-3-030-86144-5_4

and the increasing presence of AI-fuelled autonomous systems have shown that privacy concerns are insufficient: ethics and the human dignity are at stake. "Accept/ not accept" options do not satisfy our freedom of choice, and what about our individual preferences and moral views?

Autonomous machines tend to occupy the free space in a democratic society in which a human being can exercise her freedom of choice. That is the space of decisions that are left to any individuals when such decisions do not break fundamental rights and laws but are the expression of personal ethics. From the case of privacy preferences in the app domain to the more complex case of autonomous driving cars, the potential user is left unprotected and inadequate in her interaction with the digital world.

A simple system that manages a queue of users to access a service by following a by design fair ordering, e.g., first in first out, may prevent users to exchange their positions in the queue by personal choice, thus depriving users to exercise a free choice driven by her moral disposition, e.g., leave her position to an older lady.

What is considered fair by the system's developer may not match users' ethics.

The above example may seem artificial and of little importance, but it is not. In the years of digital transformation we have witnessed the side effect of increasing the rigidity of the processes implemented by digital systems beyond what the law indicated. How many times have we heard answers like "yes this could be possible, but the system does not allow it"? The above queue managing system may have associated to the ordering position a personal identifier and already made available all the personal information to the service provider. Although it may appear more complex to exchange positions, it would be not a problem for a digital system to manage the exchange. It only requires that the system is properly designed to take into consideration the user's right of choice. Overlooking this attitude in the era of autonomous technology may put at high danger our personal ethical values.

More complex interactions between systems and users shall be made possible in order to allow users' ethics to freely manifest.

However, even when such interaction is made possible, think, e.g., of the (by GDPR law mandatory in Europe) possibility to express users' consent to cookies profiling, the ways systems are presenting such interaction to the user is extremely complex and time expensive even for an expert user and often turns out in a accept/ not accept choice.

In a digital society where the relationship between citizens and machines is uneven, moral values like individuality and responsibility are at risk.

From a societal point of view, it is therefore crucial to understand which is the space of autonomy that a system can exercise without compromising laws and human rights.

Indeed, autonomous systems interact within a society, characterized by collective ethical values, with multiple and diverse users, each of them characterized by her individual moral preferences.

The European Group on Ethics in Science and New Technologies (EGE) recommends an overall rethinking of the values around which the digital society is to be structured (EGE 2018), the most important being the value of human dignity in the context of the digital society, understood as the recognition that a person is worthy of respect in her interaction with autonomous technologies. A person must be able to exercise control on information about herself and on the decisions that autonomous systems make on her behalf.

There is a general consensus about this, but legislation follows and does not prevent the problem, and it is debatable whether regulatory approaches like GDPR or others are effectively protecting the human dignity of users. Besides regulation, active approaches have been proposed in the research on AI, where systems/software developers and companies should apply ethical codes and follow guidelines for the development of trustworthy systems in order to achieve transparency and account-ability of decisions (AI HLEG 2019; EU 2020). However, despite the ideal of a human-centric AI and the recommendations to empower the users, the power and the burden to preserve the users' rights still remain in the hands of the (autonomous) systems producers.

The above-described active approaches do not guarantee our freedom of choice that is manifested by our individual preferences and moral views. Design principles for meaningful human control over AI-enabled AS are needed. Users need (digital) empowerment in order to move from passive to active actors in governing their interactions with autonomous systems, and it is necessary to define the border in the space of decisions between what the system can decide on its own and what may be controlled and possibly overridden by the user. This also means that the system shall be designed to be open to more complex interactions with its users as far as users' moral decisions are concerned.

But how to draw the border between systems' decisions and users' ones?

Reflections on digital ethics can help in this respect. Digital ethics, as introduced in Floridi (2018), is the branch of ethics that aims at formulating and supporting morally good solutions through the study of moral problems relating to personal data, (AI) algorithms, and corresponding practices and infrastructures. It identifies two separate components, hard and soft ethics. Hard ethics is the base to define and enforce values by legislation and institutional bodies, i.e., hard ethics is what makes or shapes the law and represents the values collectively accepted, e.g., GDPR in Europe.

It is insufficient, since it cannot and shall not cover all the space of ethical decisions. Soft ethics complements it by considering what ought and ought not to be done over and above the existing regulation, not against it, or despite its scope, or to change it, or to by-pass it (e.g., in terms of self-regulation).

Personal preferences fall in the scope defined by soft ethics, e.g., the varieties of privacy profiles that characterize different users. A system will implement decisions to choices that correspond to both hard and soft ethics. The producer will guarantee compliance with the hard ethics rules but who does take care, and how it can care of, the values and preferences of each person?

We claim that soft ethics can express users' moral preferences and should mold their interaction with the digital world. Empowering a person with a software technology that supports her soft ethics is the means to make her an independent and active user in/of the digital society.

Depending on the system's stakeholders, the term user includes individuals, groups, and the society as a whole.

Thus, the capability of AS to make decisions does not only need to comply with legislation but also with any users' moral preferences (including privacy) when they manifest. This leads to the challenge of dealing with moral agreements between the system's hard and soft ethics (e.g., implemented by the system producer) and the user's soft ethics in making her decisions. It is worth noticing that if the user's soft ethics does not manifest, the system will make decisions according to hard ethics and its default soft ethics, e.g., the fair ordering algorithm of our queue example.

Let us now discuss a more elaborated example that is set in the automotive domain.

1. Setting: A parking lot in a big mall.
2. Resource contention: Two autonomous connected vehicles (named A and B hereafter), with one passenger each, are competing for the same parking lot. Passenger of vehicle B has a bad health status.
3. Context: A and B are rented vehicles, therefore they are multi-user and have a default ethics that determines their decisions. The default ethics of A and B (provided by the cars' producers) are utilitarian. Thus, the cars will look for the free parking lot that is closer to the point of interest; in case of contention, the closest car gets in.
4. Action: A and B are approaching the parking lot. A is closer, therefore it would take the parking lot. However, by communicating with B, it receives the information that the passenger in B is in bad health condition. Indeed, the passenger in B, who has a tradeoff privacy disposition, has disclosed such piece of personal information. The soft ethics of the passenger in A has a generosity disposition that manifests in the presence of bad health people, and, consequently, actions are taken to leave the parking lot to B. This use case shows how personal privacy is strictly connected to ethics: by disclosing a personal piece of information like this, the bad health passenger's tradeoff privacy disposition manifests the utilitarian expectation that surrounding drivers might have generosity disposition.

This example shows something that already happens in our ordinary reality when, e.g., owners of cars display signs, e.g., baby on board, beginners at drive, or disabled on board, about the persons in the car.

If one could imagine a sort of soft ethical initial configuration step of the vehicle for a single owner, what would happen if the car is multi-owner or the business model in the automotive domain changes from proprietary to rented due to the increased autonomy? How will the passenger disclose her piece of information and inform the surrounding vehicles? And how would a passenger be able to set the soft ethical part of the autonomous vehicle decisions according to her own ethical preferences?

From a system design perspective, there is the need of a software architectural view of the digital world which decouples autonomous systems from users as independent au pair actors. Users need to be digitally empowered in order to be able to make, possibly complex, interactions with the surrounding AS through reliable (with respect to users' ethical preferences) protocols.

In the above direction, the separation between hard and soft ethics (Floridi 2018), initial results on design principles to empower the user (Autili et al. 2019) and on achieving moral agreements among autonomous stakeholders (Liao et al. 2019) can be exploited to concur at realizing the principle of human dignity as stated by the EGE.

Acknowledgements The work described in this chapter is part of the EXOSOUL project (https://exosoul.disim.univaq.it/). The author thanks all the research team for enlightening discussions and work together.

References

AI HLEG (2019) The High-Level Expert Group on Artificial Intelligence. Ethics guidelines for trustworthy AI. https://ec.europa.eu/digital-single-market/en/news/ethics-guidelines-trustworthy-ai

Autili M., et al. (2019) A Software Exoskeleton to Protect and Support Citizen's Ethics and Privacy in the Digital World. IEEE Access 7, 2019.

EGE (2018) European Group on Ethics in Science and New Technologies. Statement on artificial intelligence, robotics and autonomous systems. https://ec.europa.eu/research/ ege/pdf/ ege_ai_statement_2018.pdf.

EU (2020) European Commission. White paper on artificial intelligence, 2020. https://ec.europa.eu/info/sites/info/files/commission-white-paper-artificial-intelligence-feb2020_en.pdf

Floridi L. (2018) "Soft ethics and the governance of the digital." *Philosophy & Technology* 31(1), pp. 1-8.

Liao B., Slavkovik M., and van der Torre L. (2019) Building Jiminy Cricket: An Architecture for Moral Agreements Among Stakeholders. In Proc. of AIES19, 2019.

Part II
Participation and Democracy

The Real Cost of Surveillance Capitalism: Digital Humanism in the United States and Europe

Allison Stanger

Abstract Shoshana Zuboff's international best seller, *The Age of Surveillance Capitalism: The Fight for a Human Future at the Frontier of Power*, has rightfully alarmed citizens of free societies about the uses and misuses of their personal data. Yet the concept of surveillance capitalism, from a global perspective, ultimately obscures more than it reveals. The real threat to liberal democracies is not capitalism but the growing inequalities that corporate surveillance in its unfettered form both reveals and exacerbates. By unclearly specifying the causal mechanisms of the very real negative costs she identifies, Zuboff creates the impression that capitalism itself is the culprit, when the real source of the problem is the absence of good governance.

Although often portrayed otherwise, surveillance in and of itself is neither inherently good nor evil. Legal surveillance for national security reasons is essential in protecting the homeland. Citizens' video surveillance built the case against the incendiary January 6 attackers of Congress, just as social media played a significant role in the organization of that siege. The same social media that fuels micro-targeted advertising also played a critical role in the Black Lives Matter movement's ability to change world public opinion on the importance of racial justice in a breathtakingly short period of time.

Big Tech did not create extremism and polarization in the United States and Europe, but unfettered data harvesting has certainly undermined human values. Shoshana Zuboff's international best-seller, *The Age of Surveillance Capitalism: The Fight for a Human Future at the Frontier of Power*, has rightfully alarmed citizens of free societies about the uses and misuses of their personal data. Ahead of the curve, Europe has already taken innovative steps with the General Data Protection Regulation (GDPR) and the Digital Services Act (DSA) to contain the negative repercussions of unregulated markets. Yet the concept of surveillance capitalism, from a global perspective, ultimately obscures more than it reveals. Capitalism is not

A. Stanger (✉)
Middlebury College, Middlebury, VT, USA
e-mail: stanger@middlebury.edu

© The Author(s) 2022
H. Werthner et al. (eds.), *Perspectives on Digital Humanism*,
https://doi.org/10.1007/978-3-030-86144-5_5

the cause of surveillance, since the same insatiable drive for data exists in planned economies. China currently leads the world in AI applications, because it also leads the world in commercial and security espionage. The real threat to liberal democracies is not capitalism, as Zuboff's book title seems to imply, but the growing inequalities that corporate surveillance in its unregulated present form both reveals and exacerbates.

1 Zuboff's Argument

It is impossible to summarize the complicated argument of Zuboff's book easily. A lengthy January 29, 2021, opinion piece in the *New York Times* suggests that Zuboff's thinking has evolved since writing it. The adjective "epistemic" features prominently in this unusually lengthy essay (she uses it 36 times in her 5000 word piece), yet the word "epistemic" appears *not once* in her 702-page book.[1] Thus, this essay seems vital for understanding her current views, after decades of research, on some of the most important takeaways from her magnum opus (Zuboff 2019, 2021).

In the *New York Times* piece, Zuboff argues that the world is currently experiencing both "epistemic chaos" and an "epistemic coup." The Big Tech companies have executed a silent epistemic coup, Zuboff warns, to which we must pay close attention if we are to sustain democracy. "In an information civilization, societies are defined by questions of knowledge—how it is distributed, the authority that governs its distribution and the power that protects that authority," Zuboff writes. "Who knows? Who decides who knows? Who decides who decides who knows? Surveillance capitalists now hold the answers to each question, though we never elected them to govern. This is the essence of the epistemic coup" (Zuboff 2021).

Zuboff's epistemic coup proceeds in four stages. The first is the "appropriation of epistemic rights." The second involves a rise in epistemic inequality, "The difference between what I can know and what can be known about me." The third and present stage is one of "epistemic chaos" that is the result of prior coordinated manipulation; this is the stage where there is disagreement about the truth that cannot be bridged. The fourth is epistemic dominance, effectively the institutionalization of computational government by "private surveillance capital." The epistemic chaos reflected in the January 6 siege of the capital, according to Zuboff, was a warning shot (Zuboff 2021).

Who or what is driving this sequential epistemic nightmare? Presumably the appropriators are private surveillance capitalists, and the implication is that Zuboff is mounting a Marxist argument, as she seems to be pointing a finger squarely at capitalism itself. She also sees the CIA and NSA—i.e., government—as part of surveillance capitalism. Is this the deep state as the executive committee of the bourgeoisie? While Marx references feature prominently in Zuboff's book, they are

---

[1] These epistemic rights are apparently self-evident, since they are never defined.

not mentioned in the *New York Times* essay, suggesting that capitalism itself isn't the problem. But if the capitalist profit motive is not the problem, what is? What does any of this have to do with epistemology? Big data in and of itself is not knowledge. The interpretation of data produces knowledge rather than mere noise. Who are these knowledge creators? Is Zuboff one of them?

2 What the Metaphor of Surveillance Capitalism Obfuscates

Zuboff's concepts of surveillance capitalism and a related epistemic coup pose obstacles to understanding, for at least four reasons. First, coups, which involve the military, have clear objectives in mind. They involve an intention to seize power. The military has not sanctioned Facebook and Google's rising power. Yes, Facebook knows more about us than we know about ourselves and each other, and they cut off the communications of Donald Trump just as the military will close down airports and all communications in the aftermath of a coup. But not only is the military not involved in any of this, there is nothing to stop Donald Trump or any one of us, to communicate by alternative means. Trump can still send a mass email, and we can still delete and block, or simply migrate to another platform. Facebook and Twitter appear to have total information dominance only when we allow them to do so. We are choosing to allow companies to commodify our personal data and use us in this way.

Second, there is a difference between voluntary and involuntary surveillance, between whether the citizen volunteers the information or has it hijacked by a company or government agency. From a foreigner's perspective, the NSA conducts involuntary surveillance when it exploits the Court's interpretation of the Fourth Amendment, which prohibits unreasonable searches and seizures, as not extending to foreigners. According to existing American constitutional interpretation, American citizens have the right to privacy; non-American citizens do not. Perhaps extending Fourth Amendment protections to Europeans is a promising change to consider, but the question that rightfully then arises is why Europeans should receive a privilege that non-Europeans do not. The status quo jurisprudence has the virtue of being non-discriminatory, as it applies to all non-citizens. In addition, consumers may volunteer their personal data to Facebook in exchange for using the platform, but that initial consent could in the future be ruled involuntary when Facebook sells the data to third parties or stands as a gatekeeper to the use of other apps requiring a Facebook login.

Third, rather than government and business being co-conspirators in surveillance capitalism, as Zuboff suggests, it is possible to find numerous instances where Big Tech and governments have been at loggerheads, both within the United States and especially when one extends one's gaze beyond American borders. Twitter and Facebook banning the former American president from their platforms is the most recent example of the former. The myriad ways that the capitalist economies of continental Europe and the European Union have challenged the excesses of sur-veillance capitalism through national and EU legislation are evidence against a business-government conspiracy. If there has been an epistemic coup, as Zuboff argues, democratic governments are clearly not entirely on board.

Finally, surveillance capitalism suggests that there is something intrinsic to capitalism that is animating data collection, when this is not the case. While its economy certainly has some capitalist features, the Chinese Communist Party's interventions in economic life are incompatible with capitalism. Beijing's restructuring of Alibaba co-founder Jack Ma's corporate empire is a case in point (Zhong 2021). The China Brain Project, which involves the Chinese military, harvests data from Baidu that fuels China's controversial Social Credit System, designed to reward "pro-social" and punish "anti-social" behavior. China is also apparently interested in building expertise in American behavior modification, and Americans have willingly volunteered their personal data for Chinese use in exchange for using the popular app TikTok, which has been banned in India for national security reasons. It may not be the case, however, that securing American data to train algorithms to better manipulate or sell to Americans is necessary. A 2007 multinational study found that the OCEAN Big Five personality inventory, which was exploited so brilliantly by Cambridge Analytica to interfere in the 2016 US presidential election, is "robust across major areas of the world" (Schmitt et al. 2007).

In mischaracterizing the nature of the problem, therefore, Zuboff misses the real story. Coups are intentional, and if anything, technology companies don't want the political power that has inadvertently accrued to them through their monopoly of information. Mark Zuckerberg and other tech titans have repeatedly stated that they *want* appropriate government regulation but have been forced to do the best they can with self-regulation until government again assumes its proper role as overseer and promoter of the greater good. The inherent problems with self-regulation are obvi-ous. But it is not difficult to see that government has been slow to come to terms with the dramatic transformation of democracy's public sphere through technological innovation.

To summarize, the Big Tech companies have not orchestrated a coup; they have myopically optimized for shareholder value at the expense of civic life. They have created products for other children that they do not want their own children to use (*The Social Dilemma* 2020). Further, these companies haven't cornered the market on knowledge, as the word epistemic suggests, but on data, and data can mislead just as easily as it can inform.

3 Open vs. Closed Societies: Consider China

By failing to specify the causal mechanisms of the very real negative costs she identifies, Zuboff creates the impression that capitalism itself is the cause for the problem, when the real source of the problem is the absence of good governance. Blaming capitalism itself is misplaced, because we are in the midst of a transformation that challenges our existing cognitive capacity, with or without AI. The move to the cloud, a market Amazon is betting heavily on, only exacerbates anti-democratic trends to which democratic governments have been slow to react—but are capable of doing so.

Zuboff's bottom line, however, does highlight a looming challenge to open societies and democracy: the accelerating competition between the United States and China for supremacy in AI applications and the potential implications that contest has for inalienable rights in a liberal democracy. The Chinese regime is an example of what the philosopher Elizabeth Anderson calls private government. Private government's distinguishing feature is that it does not recognize a protected public sphere free of sanction or elite oversight (Anderson 2017, p. 37). Private government is always authoritarian, since it does not value liberal notions of democratic accountability. "Private government," Anderson writes, "is government that has arbitrary, unaccountable power over those it governs" (Anderson 2017, p. 45). The ends of communist government, Anderson continues, are neither liberty nor equality but "utilitarian progress and the perfectibility of human beings under the force of private government" (Anderson 2017, p. 62).

For Anderson, the only way to preserve and protect both equality and freedom is to make government a public affair, accountable to the governed. The transition from monarchy to liberal democracy, in this view, involved gradually replacing private government with public government. Public government utilizes the rule of law and substantive constitutional rights to advance and protect the liberties and interests of the governed rather than the governors (Anderson 2017, pp. 65–66).

Government is private in China, in contrast, because the Chinese leadership rejects the very idea that the Party's encroachment on individual rights can be inappropriate or undesirable. Speaking at the Kennedy School in February 2020, former FBI director James Comey identified this difference as the place where negotiations with China over technology transfer typically break down. The Chinese don't understand the American distinction between technology for private uses and for public uses (the latter being the potential regulable space, from an American perspective) (Comey 2020). The same refusal to distinguish between the private and public realms underlies China's one child policy and the government's current efforts to encourage Chinese single women to marry and have children. Since the very idea of a right to privacy presupposes a public-private distinction, privacy in China is easily sacrificed at the altar of national security and societal goals. Thus, there is a values alignment problem for AI applications in open societies that does not exist in China (Lanier and Weyl 2020; Stanger 2021).

The Chinese embrace of an automated world powered by statistical machine learning is at odds with the very idea of public and individual rights-based government, where all are to be equal before the law. In other words, the West cannot do what China is doing with AI without compromising core liberal democratic values. At his first Town Hall, President Biden suggested that both he and Xi understand the significance of this values gap: "The central principle of Xi Jinping is that there must be a united, tightly controlled China. And he uses his rationale for the things he does based on that. I point out to him, no American president can be sustained as a president if he doesn't reflect the values of the United States. And so the idea I'm not going to speak out against what he's doing in Hong Kong, what he's doing with the Uyghurs in western mountains of China, and Taiwan, trying to end the One-China policy by making it forceful, I said—by the way, he said he gets it" (Biden 2021).

4 What Open Societies Need to Do to Remain So

It is certainly true that the people cannot govern themselves if unable to distinguish fact from fiction. Because of the possibility of illiberal democracy (Trump is exhibit A), we should not just be interested in democracy but in the *quality* of democracy. There is a real link between liberal democracy and education, the ability to distinguish truth from lies, to respect science and free inquiry. The problem is the exploitation of personal data to change behavior, not big data itself, which can be deployed for both positive and negative ends (Guszcza et al. 2014).

The real cost of the cluster of trends in motion that Zuboff calls surveillance capitalism is increasing knowledge inequality that destabilizes liberal democracy. These growing power gaps exist at both the national and global levels. They exist between the people and elites, between the most powerful tech companies and the governments who seek to regulate them, and between the companies and their product, which is you. With the GDPR and the DSA, Europe provides a laboratory for promoting greater equality in a transformed global economy. In thinking about the future of Section 230 of the 1996 Communications Decency Act, America would do well to review the European data already in hand.

Silicon Valley's disproportionate power is not imperial, because it is not wielded via Washington; rather, Silicon Valley has recently silenced a US president. Both Europe and the United States have a shared interest in educating citizens to vote with their feet so as to level the playing field in those countries where Big Tech's impact is oversized and stifles indigenous innovation. Third-party markets in personal data can be regulated. With the new Biden administration at the helm, there is a serious opportunity for European-American collaboration on AI innovation to check rising digital authoritarianism. There has to be room for greater collaboration on products that can be customized to meet different local needs. The current American lawsuit against Facebook that charges them with illegally buying up their rivals (Instagram and WhatsApp) is also something to watch. Forty US states have filed the lawsuit, and the successful antitrust case against Microsoft in the 1990s was also a product of extensive involvement of states' attorneys general in the litigation process (Kang and Isaac 2020).

Reducing social inequality premised on knowledge inequality in the face of accelerating technological change is a shared challenge. Our most pressing problems have global dimensions, which provide fertile ground for cooperation rather than confrontation.[2] For the United States, personal data ownership, the right to be forgotten, liberal education, and insisting on greater transparency in algorithmic judgments are promising places to start (Lanier 2014; Post 2018). Both Europe and the United States need to reimagine rights-based democratic government, in which every human being is worthy of education, work, and health, for the global information age. As the March 2020 Final Report of the US National Security Commission on Artificial Intelligence writes, "We want the United States and its allies to exist in a world with a diverse set of choices in digital infrastructure, e-commerce, and social media that will not be vulnerable to authoritarian coercion and that support free speech, individual rights, privacy, and tolerance for differing views" (Schmidt et al. 2021, p. 28). This is a formidable educational and political undertaking, one best tackled collaboratively with other open societies, but it is essential if we are to build a shared future that promotes the human flourishing of all, not just knowledge elites.

References

Anderson, E. (2017) *Private government*. Princeton: Princeton University Press.

Biden, J. (2021) CNN Town Hall Meeting, February 16, 2021. Transcript available at: https://www.newsweek.com/joe-biden-cnn-town-hall-transcript-full-trump-vaccines-1569872 (Accessed: 13 April 2021)

Comey, J. (2020) Speech at Harvard Kennedy School Forum, February 24, 2020.

Guszcza, J., Schweidel, D. and Dutta, S. (2014) "The personalized and the personal: Socially responsible innovation through big data," *Deloitte Review* (14), 18 January 2014. Available at: https://www2.deloitte.com/us/en/insights/deloitte-review/issue-14/dr14-personalized-and-personal.html (Accessed: 13 April 2021)

Kang, C and Isaac, M. (2020) "US and states say Facebook illegally crushed competition," *New York Times*, December 9. Available at: https://www.nytimes.com/2020/12/09/technology/facebook-antitrust-monopoly.html?referringSource=articleShare (Accessed: 13 April 2021)

Lanier, J. (2014) *Who owns the future?* New York: Simon and Schuster.

Lanier, J. and Weyl, G. (2020). "AI is an ideology not a technology," *Wired*, March 15. Available at: https://www.wired.com/story/opinion-ai-is-an-ideology-not-a-technology/ (Accessed: 13 April 2021)

---

[2]Zuboff wrote another *New York Times* opinion piece that ran on January 24, 2021. It identified epistemic inequality as the social and political harm of greatest concern. Five days later, she published her second piece, analyzed above, targeting an epistemic coup. Taken together, the two pieces suggest very different policy remedies. See https://www.nytimes.com/2020/01/24/opinion/sunday/surveillance-capitalism.html.

Post, R. (2018) "Data privacy and dignitary privacy: Google Spain, the right to be forgotten, and the construction of the public sphere," *Duke Law Journal* (67), pp. 981-1072.

Schmidt, E. et al. (2021) *Final report*, National Security Commission on Artificial Intelligence. Available at: https://www.nscai.gov/wp-content/uploads/2021/03/Full-Report-Digital-1.pdf (Accessed: 13 April 2021)

Schmitt, D. et al. (2007) "The geographic distribution of big five personality traits," *Journal of Cross-Cultural Psychology*, vol. 38 (2), pp. 173-212. Available at: https://www.toddkshackelford.com/downloads/Schmitt-JCCP-2007.pdf (Accessed: 13 April 2021)

The Social Dilemma (2020) Directed by Jeff Orlowski (documentary). A Netflix Original.

Stanger, A. (2021) "Ethical challenges of machine learning in the US-China AI rivalry," unpublished manuscript.

Zhong, R. (2021) "Ant Group announces overhaul as China tightens its grip," *New York Times*, April 12. Available at: https://www.nytimes.com/2021/04/12/technology/ant-group-alibaba-china.html (Accessed: 13 April 2021)

Zuboff, S. (2019) *The age of surveillance capitalism: The fight for a human future at the frontier of power*. New York: Public Affairs.

Zuboff, S. (2021) "The coup we are not talking about," *New York Times*, January 29. Available at: https://www.nytimes.com/2021/01/29/opinion/sunday/facebook-surveillance-society-technology.html (Accessed: 13 April 2021)

Allison Stanger is the 2020–21 SAGE Sara Miller McCune Fellow at the Center for Advanced Study in the Behavioral Sciences, Stanford University; Leng Professor of International Politics and Economics, Middlebury College; and External Professor, Santa Fe Institute

Democratic Discourse in the Digital Public Sphere: Re-imagining Copyright Enforcement on Online Social Media Platforms

Sunimal Mendis

Abstract Within the current European Union (EU) online copyright enforcement regime—of which Article 17 of the Copyright in the Digital Single Market Directive [2019] constitutes the seminal legal provision—the role of online content-sharing service providers (OCSSPs) is limited to ensuring that copyright owners obtain fair remuneration for content shared over their platforms (role of "content distributors") and preventing unauthorized uses of copyright-protected content ("Internet police"). Neither role allows for a recognition of OCSSPs' role as facilitators of democratic discourse and the duty incumbent on them to ensure that users' freedom to engage in democratic discourse are preserved. This chapter proposes a re-imagining of the EU legal framework on online copyright enforcement—using the social planning theory of copyright law as a normative framework—to increase its fitness for preserving and promoting copyright law's democracy-enhancing function.

Online social media platforms that are open to members of the public (e.g., Facebook, Twitter, YouTube, TikTok) constitute digital spaces that provide tools and infrastructure for members of the public to dialogically interact across geographic boundaries. Given the high numbers of users they attract, the substantial amount of discourse taking place over these platforms, and its capacity to influence contemporary public opinion, it is possible to define them as a core component of the contemporary digital public sphere.

As envisioned by Habermas (1989), the digital public sphere has a key function in fostering democratic discourse which, within the deliberative democratic ideal, is crucial for furthering the democratic decision-making process. Thus although typically subject to private ownership, given their character as a component of the digital public sphere, it is necessary that the governance of online social media platforms reflects this private-public partnership and is aimed towards advancing the public interest in fostering democratic discourse.

S. Mendis (✉)
Tilburg University, Tilburg, The Netherlands
e-mail: L.G.S.Mendis@tilburguniversity.edu

© The Author(s) 2022
H. Werthner et al. (eds.), *Perspectives on Digital Humanism*,
https://doi.org/10.1007/978-3-030-86144-5_6

A precise definition of the term "democratic discourse" is difficult to come by. However, Dahlberg's concept of rational-critical citizen discourse provides a notion of public discourse that is autonomous from both state and corporate power and enables the development of public opinion that can rationally guide democratic decision-making (Dahlberg 2001). This pre-supposes the ability of members of the public to engage in autonomous self-expression (free of suppression and manipulation by corporate or state power), leading to a proliferation of diverse viewpoints that can be the subject of open, inclusive, and deliberative discussion. The value attributed to fostering democratic discourse within the European Union (EU) is underscored by the fundamental rights to freedom of expression, information, arts, and culture guaranteed by the European Convention on Human Rights (ECHR 1953) and the EU Charter of Fundamental Rights (CFR 2000).

Platforms' owners or—as they are commonly referred to in copyright law discourse—online content-sharing service providers (OCSSPs) have traditionally been viewed as "mere conduits" that provide the infrastructure required for third parties to disseminate information. However, there is increasing recognition of OCSSPs' capacity to influence and direct the discourse taking place on their platforms, particularly by means of content moderation (Pasquale 2010). Content moderation refers to the governance mechanism through which OCSSPs ensure that content shared on the platforms comply with applicable legal rules and regulatory requirements. In the context of copyright law, for instance, this would refer to OCSSPs' ability to take down or flag copyright-infringing content. Content moderation enables OCSSPs to both restrict users' ability to engage in discourse (e.g., by suppressing copyright-infringing content or blocking user accounts) and to exercise a normative influence in shaping user behavior and perceptions (e.g., by interpreting the scope of a copyright exception in their terms of service). Hence, although they lack editorial control over content shared on their platforms, OCSSPs fulfill a crucial role as curators of online discourse and as arbiters of online copyright enforcement. In executing this curatorial function, OCSSPs have the capacity to function as facilitators of democratic discourse by ensuring that their content moderation systems are designed and implemented in a manner that minimizes the negative effects of harmful discourse through the suppression of illegal (e.g., copyright-infringing) content while safeguarding users' freedom to engage with creative and cultural content in socially valuable ways, thereby promoting robust public discourse.

The creation and sharing of user-generated content (UGC) is a primary means by which users of social media platforms engage in discourse. Remix practices that involve the re-use and re-interpretation of existing informational and cultural content (e.g., texts, images, music) in creative ways for purposes of social commentary and critique (e.g., parodies, memes) typically for non-commercial purposes constitute a particularly powerful mode of dialogic interaction that enables the dissection of contemporary narratives to create new meaning by challenging established ideological assumptions and stereotypes (Peverini 2015). It is noteworthy that copyright law recognizes the importance of facilitating the re-use and re-interpretation of informational and cultural content for purposes of social commentary and critique by means of granting specific exceptions and limitations that enable the use of copyright-protected content for purposes of quotation, criticism, review, and parody. These exceptions qualify as user privileges[1] or freedoms that enable users to engage with copyright-protected content without fear of legal sanction. Thus, ensuring the effective protection of these users' freedom to engage with copyright-protected content for purposes of social commentary and critique gains paramount importance in fostering democratic discourse on social media platforms. Accordingly, the Court of Justice of the European Union (CJEU) has on multiple occasions underscored the importance of interpreting such exceptions and limitations in a manner that ensures the protection of users' freedom of expression (Article 11 CFR) and freedom of the arts (Article 13 CFR) while striking a fair balance with copyright owners' fundamental right to the protection of their intellectual property (Article 17(2) CFR).[2]

However, the existing EU law framework on copyright enforcement in the online sphere—of which Article 17 of the *Copyright in the Digital Single Market Directive* [2019] constitutes the seminal legal provision—is rooted in the utilitarian viewpoint that envisages copyright as an instrument for incentivizing the production of creative content by granting authors a means of obtaining an adequate reward for their intellectual investment. This is reflected in the underlying policy rationale for Article 17 which is bridging the "value gap" that purportedly stems from the under-compensation of copyright owners (especially in the music industry) for copyright-protected content shared by users over online content-sharing platforms. Thus, the primary goal of Article 17 is to protect the economic interests of copyright owners by preventing the unauthorized distribution of copyright-protected content over online platforms and ensuring their ability to obtain fair remuneration through licensing. Article 17 seeks to achieve this goal by imposing a heightened duty of care and an enhanced degree of liability on OCSSPs for copyright-infringing content shared by users over their platforms. Consequently, OCSSSPs are compelled to adopt a more stringent regulatory approach toward preventing the sharing of unauthorized content, for instance, by engaging in more expansive monitoring and filtering of UGC content with the aid of algorithmic content moderation tools (Frosio and Mendis 2020).

Conversely, the preservation of users' freedom is rendered peripheral to the core economic aim of ensuring fair remuneration to copyright owners. Although Article 17(7) enunciates that OCSSPs' efforts to suppress infringing content should not result in preventing the availability of non-infringing content (as in the case where the use of copyright-protected content falls within a legally granted exception), no explicit liability is imposed on OCSSPs for the wrongful suppression of legitimate uses of copyright-protected content. Neither are OCSSPs imposed with enforceable

---

[1] It is noted that in the decisions delivered in C-469/17 *Funke Medien NRW* [2019] ECLI:EU:C: 2019:623 and C-516/17 *Spiegel Online* [2019] ECLI:EU:C: 2019:625 the Court of Justice of the European Union (CJEU) interpreted the exceptions and limitations to copyright provided under Article 5 of the EU Copyright Directive (2001) as user rights as opposed to mere user privileges.

[2] For instance, Case C-201/13 Deckmyn v Vandersteen [2014] ECDR 21, C-469/17 *Funke Medien NRW* [2019] and C-516/17 Spiegel Online [2019]. *See also* C-476/17 *Pelham and Others* [2019] ECLI:EU:C: 2019:624.

obligations to safeguard users' freedom. On the other hand, Article 17(7) does underscore the importance of ensuring that users are able to rely on existing exceptions and limitations for quotation, criticism, review and parody, caricature and pastiche, but this responsibility is assigned to Member States as opposed to OCSSPs.

Thus, the online copyright enforcement regime introduced by Article 17 is skewed in favor of protecting the economic interests of copyright owners with less emphasis being placed on the protection of users' freedom to engage in democratic discourse. Pursuant to a simple "cost-benefit" analysis, it is intuitive that it would be less costly for OCSSPs to block or remove questionable content rather than to take the risk of incurring liability under Article 17. This means that OCSSPs would be incentivized to calibrate their content moderation systems to suppress *potentially* copyright-infringing content without properly analyzing the legality of that content under applicable copyright exceptions, thereby increasing the risks of "collateral censorship."

As such, within the present EU legal framework on online copyright enforcement, the role of OCSSPs is limited to ensuring that copyright owners obtain fair remuneration for content shared over their platforms (role of "content distributors") and preventing unauthorized uses of copyright-protected content ("Internet police"). Neither role allows for a recognition of OCSSPs' role as facilitators of democratic discourse and the duty incumbent on them pursuant to that role to ensure that users' freedom to engage in democratic discourse are preserved.

While acknowledging the primacy of the utilitarian-based incentive theory as the dominant narrative of contemporary EU copyright law framework, this chapter proposes a re-imagining of the EU legal and policy framework on online copyright enforcement using the social planning theory of copyright law as a parallel theoretical framework. The social planning theory as advanced in the writings of Elkin-Koren (1995), Netanel (1996), and Fisher (2001) is rooted in the ideological argument that copyright can and should be shaped to foster a just and attractive democratic culture (Fisher 2001, p. 179). While affirming the role of copyright in preserving the incentives of authors to produce and distribute creative content, the social planning theory envisions a broader purpose for copyright law in promoting the discursive foundations for democratic culture and civic association (Netanel 1996). Thus, it prescribes that protecting the interests of copyright owners must be tempered by the overarching aspiration of sustaining a participatory culture (Fisher 2001), which in turn necessitates the adequate preservation of users' freedom to engage with copyright-protected content for purposes of democratic discourse. Accordingly, the social planning theory emphasizes the need to calibrate copyright law to minimize impediments to the public's ability to engage with content in socially valuable ways while, at the same time, protecting the legitimate interests of copyright owners (Netanel 1996).

I argue that the democratic function of copyright law, as exemplified by the social planning theory, offers a normative basis for re-imagining the role of OCSSPs within EU copyright law. This would firstly entail a re-affirmation that the protection of users' freedom to benefit from copyright exceptions (particularly those exceptions that are vital for enabling democratic discourse such as quotation and parody) is central to copyright law's purpose and as such should be granted equal weight and importance as the protection of the economic rights of copyright owners. This would enable the protection of users' freedom to be re-located from the periphery of copyright law policymaking (to which it is currently relegated) to the center of the discussion. Furthermore, it would provide a normative basis for courts to engage in a more expansive teleological interpretation of copyright exceptions with a view to advancing the democracy-enhancing function of copyright law. Secondly, it would pave the way for acknowledging the potency of content moderation systems to direct and influence public discourse on social media platforms. This would provide a basis for OCSSPs to be imposed with positive obligations to ensure that content moderation systems are designed and implemented in a manner that provides adequate protection to users' freedom, thereby transforming their role from being mere "content distributors" or the "Internet police" to active partners in fostering democratic discourse in the digital sphere.

Within the present EU copyright law discourse that is grounded on the utilitarian approach, arguments for preserving and promoting democratic discourse on social media platforms are typically rooted in fundamental rights justifications, particularly the freedom of expression. Thus, fostering a participatory culture and robust dialogic interaction in the online sphere tends to be viewed as something exogenous to copyright law's objectives and, often, as something that comes into conflict with it. Espousing the social planning theory as a parallel theoretical framework would bring about a paradigm shift that enables the protection of democratic discourse to be seen as something that is endogenous—and in fact fundamental—to copyright's purpose and provide a solid normative basis for re-imagining the EU legal framework on online copyright enforcement to increase its fitness for preserving and promoting copyright law's democracy-enhancing function.

References

Dahlberg, L. (2001). The Internet and democratic discourse: Exploring the prospects of online deliberative forums extending the public sphere. *Information, Communication & Society*, 4(4), pp. 615–633.

Elkin-Koren, N. (1995). Copyright and Social Dialogue on the Information Super Highway: The Case Against Copyright Liability of Bulletin Board Operators. *Cardozo Arts & Entertainment Law Journal*, 13, pp. 346–411.

Fisher, W. (2001). Theories of Intellectual Property. In: S.R. Munzer, ed., *New Essays in the Legal and Political Theory of Property*. Cambridge: CUP, pp. 168–199.

Frosio, G. and Mendis, S. (2020). Monitoring and Filtering: European Reform or Global Trend?'. In: G. Frosio ed., *The Oxford Handbook of Intermediary Liability*. Oxford: OUP.

Habermas, J. (1989). *The Structural Transformation of the Public Sphere: An Inquiry into a Category of Bourgeois Society.* Translated by Thomas Burger. Cambridge: MIT.

Netanel, N.W. (1996). Copyright in a Democratic Civil Society. *Yale Law Journal,* 106, pp. 283–387.

Pasquale, F. (2010). Beyond Innovation and Competition: The Need for Qualified Transparency in Internet Intermediaries. *Northwestern University Law Review,* 104 (1), pp. 105–173.

Peverini, P. (2015). Remix Practices and Activism. In: E. Navas, O. Gallagher, x. burrough, eds., *The Routledge Companion to Remix Studies.* New York: Routledge, pp. 333–345.

The Internet Is Dead: Long Live the Internet

George Zarkadakis

Abstract Social exclusion, data exploitation, surveillance, and economic inequality on the web are mainly technological problems. The current web of centralized social media clouds delivers by design a winner-takes-all digital economy that stifles innovation and exacerbates power asymmetries between citizens, governments, and technology oligopolies. To fix the digital economy, we need a new, decentralized web where citizens are empowered to own their data, participate in disintermediated peer-to-peer marketplaces, and influence policy-making decisions by means of innovative applications of participatory and deliberative democracy. By reimagining "web 3.0" as a cloud commonwealth of networked virtual machines leveraging blockchains and sharing code, it is possible to design new digital business models where all stakeholders and participants, including users, can share the bounty of the Fourth Industrial Revolution fairly.

The internet is almost 50 years old.[1] It has transformed our world and has provided new and exciting opportunities to business, society, science, and individuals; but it has also ushered an era of greater inequality, surveillance, exclusion, and injustice. The first iteration of the internet ("web 1.0") was a network of organizational servers where individual PCs were connected intermittently via dial-up modems. This evolved into the current mobile internet ("web 2.0") that consists of many centralized social-network clouds that suck user data like black holes. As such, web 2.0 has enabled the land-grabbing business models of the Big Tech oligopolies. An unfair

---

Why do we need a new internet to democratize the digital economy and enhance participatory governance?

---

[1] https://www.usg.edu/galileo/skills/unit07/internet07_02.phtml#:~:text=January%201%2C%201983%20is%20considered,to%20communicate%20with%20each%20other.&text = ARPANET %20and%20the%20Defense%20Data,the%20birth%20of%20the%20Internet

---

G. Zarkadakis (✉)
Atlantic Council, Zug, Switzerland
e-mail: gzarkadakis@atlanticcouncil.org

H. Werthner et al. (eds.), *Perspectives on Digital Humanism*,
https://doi.org/10.1007/978-3-030-86144-5_7

distribution of power has thus emerged whereby our data are gathered, analyzed, and monetized by private companies while we accept cookies in a hurry. Today's digital economy is a rentier's wet dream come true. We have become serfs in the digital fiefdoms of the techno-oligarchs, tenants, instead of co-owners, of the enormous economic value that we generate through our digital avatars. Add the march of AI systems that automate our jobs, and what you get is the social contract of liberal democracy being torn to pieces. If we continue with business as usual, the future will be one of massive unemployment, dislocation, and the end of dreams. The Global Financial Crisis offered us a glimpse of what that means: populism, mistrust in democracy, conspiracy theories, polarization, hate, racism, and a dark replay of the 1930s. The COVID-19 pandemic has further exacerbated the political and economic asymmetries of a digital economy that works only for the few. "Working from home" sounds good until you realize that your job can now be outsourced anywhere in the world, at a much lower cost. Virtualization of work equals labor arbitrage enabled by web 2.0 and Zoom calls.

1 Perils of an Ornithopter Approach

Faced with the danger of their historical obliteration, democracies are gearing up for a fight. Proposals range from breaking up the tech oligopolies, taxing them more, expanding welfare to every citizen via a universal basic income, and enacting stricter data privacy laws. The defense strategy has a noble purpose: to reduce the power and influence of the techno-oligarchs.

But the usual means of defending democracy by legislating and regulating are insufficient and inefficient to deal with the magnitude and nature of this particular problem. Instead of a viable and sustainable solution, they will create new bottlenecks, more paperwork, more centralized control, and more loopholes to be exploited by influential and well-funded lobbies. At the end, regulation shifts power to governments, not citizens. We, the citizens, will be replacing one master with another. The strategy of increasing the role of the State in order to deal with inequality in the digital economy is wrong and will fail. Our problem is technological, not regulatory. Like the early aviators, we are trying to leave the ground using wings that cannot fly. And just as you cannot regulate an ornithopter to reach the stratosphere, so it is with the current internet: to make the digital economy fairer, trustworthy, and more inclusive, we need a different technology, another kind of internet.

But what would that "alternate" internet look like? And what should its fundamental building blocks be? Perhaps the best way to think about these questions is to begin by asking what is wrong with the current technology. I would like to argue that there are three main problem areas that we need to focus in order to reinvent the internet: data ownership (and its necessary corollary, digital identity), security, and disintermediation. Let's take those areas in turn and examine them further.

2 Data Ownership and the Need of a Digital Identity

Data ownership is perhaps the biggest problem area of all. Ownership goes beyond self-sovereignty. It suggests property rights for the data, not just the right to allow permission for their use. We need to own our personal data as well as the data that we generate through our social interactions, actions, and choices. Our data is the most valuable resource in the digital economy. They are powering the AI algorithms that animate the wheels of the digital industries. When those algorithms finally replace us in the workplace, our data will be the only valuable resource for which we can legitimately claim a share in the bounty of the Fourth Industrial Revolution. Many initiatives, such as Sir Tim Berners-Lee's Inrupt project (Lohr 2021), are trying to work around the current web and provide ways for some degree of data ownership while building on existing web standards. However, by addressing only one of the three problem areas, they are only partial solutions. A more radical approach is necessary by establishing immutable, verifiable, digital identity systems as secure and trusted ways to identify every actor connected on the internet, including humans, appliances, sensors, robots, AIs, etc. Data generated by an actor would thus be associated with their digital identity and thus establish ownership rights. So, if I am a denizen of the alternate internet, I can decide what pieces of my data I will make available, to whom, and under what conditions. For example, if I need to interact with an application that requires my age, I will only allow that piece of data to be read by the application and nothing else. I can also decide on a price for giving access to my data to a third party, say to an advertising agency or a pharmaceutical company wishing to use my health records for medical research. Or I can decide to join a data cooperative, or a data trust (Zarkadakis 2020a), and pool my data with other people's data, include perhaps data from smart home appliances and smart city sensors, and thus exponentially increase the collective value of the "shared" data value chain. Digital identities can enable auditable data ownership from which we could engineer an equitable income for citizens in an AI-powered economy of material abundance. As we look for sustainable and meaningful funding for a universal basic income, data ownership based on digital identity may be the key solution. A "universal basic income" that is funded by economic activity would relieve governments from having to excessively tax and borrow. More importantly perhaps, it would be income earned, not "handed out," and as such would uphold – rather than demean – human dignity and self-respect.

3 Security

The current internet is highly susceptible to cyberattacks – such as Denial of Service (DoS) – because its architecture requires applications to be centralized and their data to be relayed via central servers, rather than exchanged directly between devices. Servers are thus single points of failure and attack that are candy for malevolent hackers. Centralization of data in the current internet is also the cause of data breaches, regulatory fines, reputational risk, and consumer mistrust. Given these inherent shortcomings, the cyber security arms race is forever unwinnable. In the alternate internet, applications should be peer-to-peer, decentralized, and communicate directly with each other. They should run on virtual machines that use an operating system and communication protocol built on top of the fundamental TCP/IP protocol and run on the individual device, say a smartphone. If they fail, or are attacked, the damage will be minimal and restricted rather than spreading across the whole of the network. Moreover, such an operating system could render that alternate internet as a single, global computer, made up of billions of nodes. It will be the realization of the original internet dream: a completely decentralized and secure global network of trust, where there can be no surveillance.

4 Disintermediation

Security is not the only negative outcome of the centralized nature of web 2.0. All services are currently intermediated by default, and that includes essential services such as data storage and computing resources. It is the reason why only four companies control 70% of the world's cloud infrastructure (Cohen 2021). Regardless of what one may think of Donald Trump, the fact that Twitter, a private company, could silence a sitting US President ought to give pause to anyone who cares about free speech. Like Twitter, many other private social media platforms such as Facebook and YouTube have assumed the high office of unelected arbitrators of what is true and permissible, replacing the role of legislatures, courts, and governments. Intermediation is also responsible for the fact that internet content can be monetized almost exclusively using an advertising business model, which is what social media platforms are exploiting in order to make their billions. If you are a content creator today, you need to satisfy advertisers with high numbers of followers and hits, which impacts both what content you can create and how you deliver it. The tiny minority of individual content creators who get this combination right and manage to earn some meaningful income from their work are then subject to the whims of the social media platforms that host their content. We must disintermediate and decentralize the internet if we want human creativity and innovation to flourish. In the alternate internet, content creators do not need the advertising industry to earn income; they can monetize their content in disintermediated, peer-to-peer markets, through micropayments paid to them directly by the consumers of their content. Moreover, infrastructure can also be disintermediated, and every node in the network can provide data and computing services and resources. A decentralized internet at global scale will provide new sources of income for billions of people. Just imagine anyone connecting to that internet via their smartphone or laptop and making available data storage and computing as part of a "cloud commonwealth."

5 The Rise of a New, Decentralized Web

We are already witnessing the dawn of the new internet, often referred to as "web 3.0." Distributed ledger technologies are providing new ways to establish disintermediated trust in peer-to-peer marketplaces, as well as new ways to reimagine money and financial assets. Smart contracts automate transactions and providence auditing on supply chains, banking, and insurance, while non-fungible tokens (NFTs) are transforming the internet of information into the internet of assets and enable content creators and digital artists to sell their creations, just as they would if they were made of atoms instead of bits.

In the web 3.0, individual users are connected directly, peer-to-peer, without centralized clouds. They may exchange content and other digital assets as NFTs that are executable applications. To achieve this, we must empower users with personal cloud computers (PC2), i.e., software-defined personal computers, to store and process their data. When users are ready to swap or trade their data, they can compile data into NFT capsules (encrypted, self-extracting, self-executing programs that encapsulate the data), much like people generate PDF files from Word documents today. And they will then share those capsules on a content delivering network running on a blockchain and verified by miners, instead of centralized cloud servers. The web 3.0 will, in effect, be a peer-to-peer web of personal cloud computers.

Many engineers and developers are already working on realizing the new internet. For example, the Elastos Foundation is developing a full range of tools and capabilities for web 3.0 application developers, so they may begin to code decentralized applications on the new web. All this is happening on open-source platforms in the true spirit of sharing knowledge and collaborating on projects. Which is exactly what is different in the mindset of the new digital world: success, innovation, and economic value can be derived from collaboration, not just competition, and from co-creating equitable ecosystems, not enforcing inequitable oligopolies.

6 Participatory Public and Private Governance

There is an additional prize to be won from transitioning from web 2.0 to web 3.0. A decentralized web based on digital identity, data ownership, and a peer-to-peer cloud commonwealth will bring new possibilities for digital business models that are inclusive and democratically governed, much like cooperatives of mutual ownership companies. Ethereum was first to experiment with "decentralized autonomous organizations" (DAOs) that can blend direct and representative democracy in private governance. However, similar models of democratic digital governance can also be adopted in the public domain by a city, a county, a region, or a nation state or in circumstances where citizens need to manage commons (Zarkadakis 2020b, p. 140).

In a society where every citizen has a verifiable and decentralized digital identity, there can be no surveillance – state or private – only liberty and personal responsibility. Citizens can freely associate in the cyberspace, transact, create, innovate, debate, learn, and self-develop. Because their identity is verifiable and trusted, they are incentivized toward socially responsible behavior and consensus. The decentralized web 3.0 can be the foundation of a truly digital "polis" in cyberspace. In such a democratic polis, free citizens own their data and contribute through their interactions, knowledge, skills, networks, and choices to the creation of economic value in inclusive digital platforms, value that is then shared fairly among all those who have contributed on the basis of their contribution. Moreover, participatory forms of public governance can be enacted by institutionalizing citizen assemblies in liberal democracies and include citizen views and recommendations in the policy-making processes of legislatures. The future does not have to be a dystopia of oppression, surveillance, endemic penury, and dependency on central government control and welfare. A new internet of equitable opportunities can help us overcome the dire straits of societal polarization and injustice and let democracy, freedom, and liberty reclaim their civilizing influence on human nature.

References

Cohen J. (2021) 'Four Companies control 67% of the world's cloud infrastructure', *PC Magazine*, [online]. Available at: https://uk.pcmag.com/old-cloud-infrastructure/131713/four-companies-control-67-of-the-worlds-cloud-infrastructure (Accessed: 18 May 2021)

Lohr, S. (2021) 'He created the internet. Now he's out to remake the digital world', *New York Times*, [online]. Available at: https://www.nytimes.com/2021/01/10/technology/tim-berners-lee-privacy-internet.html (Accessed: 18 May 2021)

Zarkadakis, G. (2020a) 'Data Trusts could be the key to better AI', *Harvard Business Review*, [online]. Available at: https://hbr.org/2020/11/data-trusts-could-be-the-key-to-better-ai (Accessed: 18 May 2021)

Zarkadakis, G. (2020b) Cyber Republic: reinventing democracy in the age of intelligent machines, Cambridge: MIT Press.

Return to Freedom: Governance of Fair Innovation Ecosystems

Hans Akkermans, Jaap Gordijn, and Anna Bon

Abstract The Vienna Manifesto on Digital Humanism attaches great importance to the innovation processes shaping the digital society. The digital humanism question we pose in this chapter is: if innovation is a shaping force, can it itself be shaped by humans and based on human values of a just and democratic society? Nowadays, innovation is commonly theorized in policy and academic research in terms of ecosystems. Although this framing makes room for multiple stakeholders and their interaction, it is limited as it still positions innovation as a natural process. Thus, it underplays the human value and societal design dimensions of technosocial innovation. We discuss some ideas and proposals for the governance of digital innovation ecosystems such that they are fair and equitable. Design-for-fairness has as its basis a just and democratic *societal* conception of freedom.

H. Akkermans (✉)
w4ra.org, Vrije Universiteit Amsterdam, Amsterdam, The Netherlands

University for Development Studies UDS, Tamale, Ghana
e-mail: Hans.Akkermans@akmc.nl

J. Gordijn
Vrije Universiteit Amsterdam, Amsterdam, The Netherlands

The Value Engineers BV, Soest, The Netherlands
e-mail: jaap@thevalueengineers.nl

A. Bon
w4ra.org, Vrije Universiteit Amsterdam, Amsterdam, The Netherlands
e-mail: a.bon@vu.nl

H. Werthner et al. (eds.), *Perspectives on Digital Humanism*,
https://doi.org/10.1007/978-3-030-86144-5_8

1 The Vienna Manifesto and Innovation

The Vienna Manifesto on Digital Humanism[1] opens by quoting Tim Berners-Lee, the inventor of the World Wide Web, in that "the system is failing" (Berners-Lee 2018). Next, it states that "while digitalization opens unprecedented opportunities, it also raises serious concerns. (...) Digital technologies are disrupting societies and questioning our understanding of what it means to be human. The stakes are high and the challenge of building a just and democratic society with humans at the center of technological progress needs to be addressed with determination as well as scientific ingenuity. Technological innovation demands social innovation, and social innovation requires broad societal engagement."

The Vienna Manifesto emphasizes the importance of innovation processes, as innovation is seen as shaping the emerging digital society. A humanist key question is then: if innovation is a shaping force, can it itself be shaped for the purpose of a more just and democratic society? If so, how?

2 Innovation Ecosystems

The traditional policy view on innovation that has been dominant for decades casts innovation foremost in terms of "invention" and subsequent "adoption" and spread of an innovative technology. The early editions of Everett M. Rogers' (2003) highly influential text *Diffusion of Innovations* reflect this view. The process of innovation is captured in terms of a metaphor borrowed from physics. Diffusion is interpreted (certainly where it concerns the received view in high-level policy making) as a relatively deterministic, mechanistic, and unidirectional phenomenon. The same physics metaphor also serves to establish order (not to say: hierarchy) in the process of research, starting from fundamental research, then applied research, to strategic research and, ultimately, technology development.

In recent years, it has become mainstream to frame the innovation process in the different terms of *ecosystems*, both in academic literature (Oh et al. 2016) and in policy (European Union 2020). This move embodies a significant change from the older policy framing of innovation. It has a clear metaphorical nature as well, however, borrowed not from physics but from biology. This change of metaphor has important consequences in several ways.

First, the process image, or high-level empirical model, of innovation changes. Rather than a mechanistic process of diffusion (with the famous "S-curve" of adoption[2]), it posits an interactive dynamic of multiple "species," i.e., the various

---

[1] https://dighum.ec.tuwien.ac.at/dighum-manifesto/ (May 2019)

[2] The S-curve (Rogers 2003, Ch. 7) refers to the S-shaped cumulative distribution function of innovation adoption. It may be mathematically derived from a very simple imitation model for the spread of an innovation within a population or market.

key actors and stakeholders in the innovation environment. This is commonly phrased as a *coevolution*, a notion also prominent in the Vienna Manifesto (cf. also Lee 2020; Nowotny et al. 2001). It is furthermore common to find process analyses not in terms of straightforward one-dimensional diffusion, but instead of nonlinear *complex adaptive systems*; see, for example, Rogers et al. (2005) and Bon (2020).

Second, the new metaphor of ecosystems is a significant break also in the policy sense. It points to simultaneous competitive as well as collaborative relationships in innovation, in contrast with neoliberal free market ideologies that only can see enterprise competition within their horizon. It acknowledges that innovation is a multi-actor process that is non-deterministic and coevolutionary. It furthermore permits a different view on who are the actual stakeholders in play. It changes the view on the role of government as enabler of innovation, but it also changes and extends the role of *civil society* and other players that have hitherto often been ignored or downplayed. The latter point has been made particularly explicit in science-and-society and science policy literature (Gibbons et al. 1994; Etzkowitz and Leidesdorff 2000; Nowotny et al. 2001; Carayannis and Campbell 2012) on new modes of knowledge production and the triple/quadruple university-industry-government-civil society *helix* organization of innovation in nations, regions, and ("smart") cities.

Nevertheless, the ecosystems metaphor as a way to understand innovation has important limitations. Although more flexible and open-ended than older physics metaphors, it still frames innovation as some kind of natural process (but now "ecological" rather than physics-mechanistic), which on its turn carries with it the (invalid) suggestion that its course is out of human hands and beyond human control.

Indeed, business and management literature on innovation has difficulty acknowledging the implication that, as a result of the fact that humans are part of the ecology as actors and stakeholders, innovation is *human-designable* (at least to some extent). It is very weak in discussing key *normative* aspects of innovation ecosystems. Who stands to benefit from (disruptive) innovation, and why? Who is in control, and for what purposes? These are important and unavoidable matters (in a democratic debate, that is) that are discussed in literature on social innovation, but typically from disciplines other than economics and business research, witness, for example (Manzini 2015). In other words, the innovation ecosystem concept needs to be *humanized*, and this can be achieved by explicating the *governance* dimension of digital technologies.

3   Governance: Ecosystems That Are Fair

As a corrective to the "failing of the system," Berners-Lee (2018) calls for a "re-decentralization" of the Web. In a coevolutionary view, this may involve both technology (e.g., SOLID[3]) and non-technology societal actions. Jairam et al. (2021) make an explicit distinction between a technology and how it is controlled, pointing out that the technology level and its governance level can have very different characteristics. For example, the Big Tech platforms rely on network "decentralized" technologies, but their governance level is in contrast strongly centralized, even monopolistic. These authors investigate blockchain technologies (such as Bitcoin, Corda, Ethereum, Tezos) and their industrial applications (e.g., smart energy scenarios such as peer-to-peer sustainable energy trading). They show that also here many forms of governance exist from highly centralized to decentralized (and often opaque).

The focus of this work is on the question how the *governance* of technologies can be decentralized.[4] To this end, these authors introduce the notion of *fair* innovation ecosystems and propose a set of design principles for fair and equitable ecosystems. Decentralized ecosystems, as a realistic alternative for the Big Tech platforms, have a fair distribution of governance power, whereby fairness is defined along the following lines (Jairam et al. 2021):

(a) **Participation.** Fair governance ensures active involvement in the decision-making process of all who are affected and other parties with an interest at stake. It includes all participants interacting through direct or representative democracy. Participants should be able to do so in an unconstrained and truthful manner, and they should be well informed and organized so as to participate fruitfully and constructively.

(b) **Rule of law**. *Equity*: all participants have legitimate opportunities to improve or maintain their well-being. Agreed-upon legal rules and frameworks, with underlying democratic principles, are enforced impartially while guaranteeing the rights of people; no participant is above the rule of law.

(c) **Effectiveness and efficiency.** Fair governance fulfils societal needs by incorporating effectiveness while utilizing the available resources efficiently. Effective governance ensures that the different governance actors meet societal needs. Fully utilizing resources, without being wasted or underutilized, ensures efficient governance.

(d) **Transparency.** Information on matters that affect participants must be freely available and accessible. The decision-making process is performed in a manner that is clear for all by following rules and regulations. Transparency also

---

[3] SOLID is a web-decentralization project led by Berners-Lee, aiming at developing a technology platform for Social Linked Data applications that are completely decentralized and fully under users' control (https://inrupt.com/solid/).

[4] The importance of good governance is explicitly recognized in the United Nations' Sustainable Development Goals (SDGs) and is the core topic of SDG 16.

includes that enough relevant information is provided and presented in easy to understand forms or media.

(e) **Responsiveness.** A responsive fair governance structure reacts appropriately and within a reasonable time frame toward its participants. This responsiveness stimulates participants to take part in the governance process.

(f) **Consensus-oriented**. Fair governance considers the different participants' viewpoints and interests before decisions are made and implemented. Such governance is defined as consensus-oriented because it aims to achieve a broad community consensus. In order to reach this wide consensus, a firm mediation structure, without any bias toward participants, should be in place.

(g) **Accountability**. Accountability is defined as responsibility or answerability for one's actions. Decision-makers, whether internal or external, are responsible for those who are affected by their actions or decisions. These decision-makers are morally or legally bound to clarify and be answerable for the implications and selected actions made on behalf of the community.

Proposals such as these lay out a strong program for *design-for-fairness* of digital technology and society governance. Fairness includes both process and outcome aspects. As we will see below, interesting and informative precursors exist also in the non-digital society, including its ecosystems.

4   Governance and Conceptions of Freedom

Vardi (2018) attributes the failing of the Internet system to a naive "hippie" notion of information freedom.[5] In his view, information has as a result become a "commons, an unregulated shared public resource" which is subject to "The Tragedy of the Commons" (Hardin 1968). Hardin's view was that commons governance of shared resources is inevitably doomed to fail, leaving as alternatives only market and state forms of governance. He derived this from the neoclassical economics theoretical assumption that humans act as rational self-interested individual agents. His anti-collective arrangement argument was welcomed by neoliberal economists who employed it to promote their ideas about free markets as key governance mechanism.[6]

---

[5] This led to a lot of debate in the Communications of the ACM. In light of the discussion above and in the remainder of this article, one may perhaps say that hippie naiveté is in assuming that a decentralized technology effortlessly leads to a governance regime that is similarly decentralized. Quod non. This technology-driven mistake is perhaps more understandable upon realizing that an earlier generation of scientists concerned about societal impacts of science were dealing with highly centralized technologies such as the *atom bomb*. See, e.g., Bernal (1939, 1958), physics professor at Birkbeck College in London and a founding father of the field now known as Science, Technology, and Society (STS).

[6] An interesting irony here is that Hardin's article has generally been received as supporting free market ideas, but Hardin was in fact writing about overpopulation and argued for the need of state

Hardin's argument was a general theoretical one. Ostrom (1990, 2010), however, deconstructed and dismantled it in an evidence-based way, through a large international set of detailed empirical case studies and extensive field research.[7] Her work makes clear that successful commons are widespread but are *not at all* "unregulated" (or free) as a shared resource. Generally, they are characterized by governance arrangements that consist of a complex array of participatory and "grassroots" democratic agreements, possibly mixed with market mechanisms as well as forms of state regulation. Ostrom's work gave rise to a theory of what she calls "polycentric governance," formulating a set of general conditions and design principles for commons-type arrangements to be successful. There are many successful and long-standing commons also in the digital world. Although due attention should be paid to the fact that digital resources have important differences from natural resources, there are interesting parallels with proposals such as those above regarding the governance of digital technology networks.

It is intriguing to observe that in virtually all discussions of governance issues, a concept of freedom is involved, although different and even conflicting ones, and often hidden in the background.[8] Following De Dijn (2020), a prevalent conception of freedom today, adhered to by neoliberals, free marketeers, and libertarians, is that of limited state power. She describes this as a major and deliberate break with much older conceptions of freedom as developed in the Humanism and Enlightenment periods, where freedom is a collective concept and lies in the ability by the people to exercise control over the way in which they are governed – at root a democratic and participatory conception of freedom. In contrast, she traces back the leave-me-alone, I-want-to-do-what-I-like individualized conceptions of freedom to the antidemocratic and counterrevolutionary forces of the seventeenth and eighteenth centuries.[9]

A neoliberal conception of freedom reduces humans to individual, self-interested, utility-maximizing agents "freely" buying on a market. It is very much a consumptive and consumerist notion: market agents acquiring and consuming services on digital platforms. This neoliberal "the-world-is-flat" notion of freedom is indeed universal ("global") but in a fully undifferentiated and uniform ("flat") way. In contrast, the societal conception of freedom pointed at here is a *productive* notion: it is one of *citizenship* that co-creates the society we (hope to) live in. It is cosmopolitan but acknowledges that freedom is contextualized (Harvey 2009; Stuurman, 2017), with due recognition of the many different and overlapping

---

coercion, even to the point that he supported China's one-child policy. In contemporary digital society terms, he was arguing not for surveillance capitalism, but for the surveillance state.

[7] Elinor Ostrom received the Nobel Prize for Economics for this work in 2009. Not only was she the first woman to receive this prize, she was a political scientist rather than economist, leading to surprise in some economist quarters.

[8] As an interesting global example, Sen (1999) describes *Development as Freedom*. Chapter 5 of his book in particular displays that the underlying conception of freedom boils down to a neoliberal market one.

[9] It is tempting to add that the Big Tech power monopolies of today demonstrate that the neoliberal conception of freedom *itself* turns out to be Hayek's "road to serfdom."

spheres and networks of human activities and relationships – including from the standpoint of the individual and their identity.

In the digital society, proper value-based digital governance (European Union 2020) is a return to freedom: the democratic and participatory freedom of Humanism and Enlightenment. Science and innovation policy has again to move forward, from the ecosystem helix frame to a much more inclusive policy of *fair* digital ecosystems. It is today's urgent task to redesign freedom in a value-based way and put it into action for a human future of our digital society.

References

Bernal, J.D. (1939) The Social Function of Science. London, UK: Routledge.

Bernal, J.D. (1958) *World Without War*. London, UK: Routledge & Kegan Paul. 2nd edn 1961. ISBN 978-0-429-28245-4

Berners-Lee, T. (2018) ACM Turing Award Lecture, given at the 10th ACM Web Science Conference on 29 May 2018 in Amsterdam. The video of the Turing Award Lecture is available at the acm.org website: https://amturing.acm.org/vp/berners-lee_8087960.cfm. The Turing Award is considered to be the Nobel Prize for Informatics.

Bon, A. (2020) *Intervention or Collaboration? Redesigning Information and Communication Technologies for Development.* Amsterdam, The Netherlands: Pangea. ISBN 9789078289258. Open Access pdf version https://w4ra.org/publications/

Carayannis E.G. and Campbell, D.F.J. (2012) *Mode 3 Knowledge Production in Quadruple Helix Innovation Systems.* New York, NY, USA: Springer. SpringerBriefs in Business 7. ISBN 9781461420613

De Dijn, A. (2020) Freedom: An Unruly History. Cambridge, MA, USA: Harvard University Press. ISBN 9780674988330

Etzkowitz, H. and Leidesdorff, L. (2000) The Dynamics of Innovation: from National Systems and "Mode 2" to a Triple Helix of University-Industry-Government Relations. Research Policy Vol. 29, pp. 109-123

European Union (2020) *Berlin Declaration on Digital Society and Value-Based Digital Government.* Signed at the ministerial meeting of the Council of the European Union on 8 December 2020. https://ec.europa.eu/isa2/sites/default/files/cdr_20201207_eu2020_berlin_declaration_on_digital_society_and_value-based_digital_government_.pdf

Gibbons, M., Limoges, C., Nowotny, H., Schwartzman, S., Scott, P. and Trow, M. (1994) The New Production of Knowledge: The Dynamics of Science and Research in Contemporary Societies. London, UK: Sage. ISBN 0-8039-7794-8

Hardin, G. (1968) The Tragedy of the Commons. *Science* Vol. 162 No. 3859 (13 December 1968), pp. 1243-1248

Harvey, D. (2009) Cosmopolitanism and the Geographies of Freedom. New York, NY, USA: Columbia University Press. ISBN 9780231148467

Jairam, S., Gordijn, J., Torres, I., Kaya, F. and Makkes, M. (2021) A Decentralized Fair Governance Model for Permissionless Blockchain Systems. In *Proceedings of the International Workshop on Value Modelling and Business Ontologies* (VMBO 2021), Bolzano, Italy, March 4-5, 2021. http://ceur-ws.org/Vol-2835/paper3.pdf

Lee, E.A. (2020) The Coevolution: The Entwined Futures of Humans and Machines. Cambridge, MA, USA: MIT Press. ISBN 9780262043939

Manzini, E. (2015) Design, When Everybody Designs: An Introduction to Design for Social Innovation. Cambridge, MA, USA: MIT Press. ISBN 9780262028608

Nowotny, H., Scott, P. and Gibbons, P. (2001) Re-Thinking Science. Cambridge, UK: Polity Press. ISBN 0-7456-2608-4

Oh, D.-S., Phillips, F., Park, S., Lee, E. (2016) Innovation Ecosystems: A Critical Examination. Technovation Vol. 54, pp. 1-6

Ostrom, E. (1990) Governing the Commons: The Evolution of Institutions for Collective Action. Cambridge, UK: Cambridge University Press. ISBN 9780521405997

Ostrom, E. (2010) Beyond Markets and States: Polycentric Governance of Complex Economic Systems. American Economic Review Vol. 100, No. 3 (June 2010) pp. 641–672. Revised version of the Nobel Prize Lecture, 2009.

Rogers, E. M. (2003) Diffusion of Innovations. 5th edn. New York, NY, USA: Free Press

Rogers, E. M., Medina, U. E., Rivera, M. A, Wiley, C. J. (2005) Complex Adaptive Systems and the Diffusion of Innovations. *The Innovation Journal* Vol. 10 (No. 3), article 3, pp. 1-25

Sen, A. (1999) Development as Freedom. Oxford, UK: Oxford University Press. ISBN 9780192893307

Stuurman, S. (2017) The Invention of Humanity - Equality and Cultural Difference in World History. Cambridge, MA, USA: Harvard University Press. ISBN 9780674971967

Vardi, M.Y. (2018) How the Hippies Destroyed the Internet. *Communications of the ACM* Vol. 61 (No. 7, July 2018), p. 9

Decolonizing Technology and Society: A Perspective from the Global South

Anna Bon, Francis Dittoh, Gossa Lô, Mónica Pini, Robert Bwana, Cheah WaiShiang, Narayanan Kulathuramaiyer, and André Baart

Abstract Despite the large impact of digital technology on the lives and future of *all* people on the planet, many people, especially from the Global South, are not included in the debates about the future of the digital society. This inequality is a systemic problem which has roots in the real world. We refer to this problem as "digital coloniality." We argue that to achieve a more equitable and inclusive global digital society, active involvement of stakeholders from poor regions of the world as co-researchers, co-creators, and co-designers of technology is required. We briefly discuss a few collaborative, community-oriented technology development projects as examples of transdisciplinary knowledge production and action research for a more inclusive digital society.

A. Bon (✉)
w4ra.org, Vrije Universiteit Amsterdam, Amsterdam, The Netherlands
e-mail: a.bon@vu.nl

F. Dittoh
University for Development Studies UDS, Tamale, Ghana
e-mail: fdittoh@uds.edu.gh

G. Lô · A. Baart
Bolesian BV, Utrecht, The Netherlands
e-mail: gossalo@bolesian.ai; andre@andrebaart.nl

M. Pini
Universidad San Martín, Buenos Aires, Argentina
e-mail: mpini@unsam.edu.ar

R. Bwana
University of Amsterdam, Amsterdam, The Netherlands
e-mail: r.m.bwana@uva.nl

C. WaiShiang · N. Kulathuramaiyer
University Malaysia, Sarawak, Malaysia
e-mail: wscheah@unimas.my; nara@unimas.my

© The Author(s) 2022
H. Werthner et al. (eds.), *Perspectives on Digital Humanism*,
https://doi.org/10.1007/978-3-030-86144-5_9

1   Inclusion, Coloniality, and the Digital Society

People from poor environments, e.g., in the Global South, are not often included in debates about the digital society. This is surprising, as impacts from digital technologies do have far-reaching consequences for their lives and future. Nowadays, the rapid co-evolution of society and technology is calling for reflection, deliberation, and responsible action. Scientists are posing the question: "Are we humans defining technology or is technology defining us?" (Lee 2020), but who are "we" in this question? Who is defining technology, and who has the knowledge, the assets, and the decision-making power?

A way to understand the impacts of digital transformation for people in the Global South is to observe the digital society through a decolonial lens. This helps to understand the, often tacit, patterns of power in the social and technological fabric. If we consider the digital society to be an image of the physical world, it will have inherited, along with other aspects, historical patterns of inequality. These patterns are referred to as "coloniality" (Mendoza 2021, pp. 46–54; Mignolo and Walsh 2018, pp. 1–12; Quijano 2016, pp. 15–18).

At the moment of writing, about three billion people in the world are unconnected from the digital society – a phenomenon often called the digital divide – but this number is rapidly decreasing. Being connected, particularly through the Internet and Web, is generally seen as the key to a better life. With the breathtaking pace in which the Internet is rolled out even in remote corners of the world, universal connectivity, with full endorsement of the United Nations,[1] might well soon be completed. The follow-on question is: will omnipresent connectivity bring social justice, equality, and a more sustainable and prosperous world closer to all?

The World Wide Web, the backbone of the digital society, was designed, according to its inventor Tim Berners-Lee, as "an open platform that would allow everyone, everywhere to share information, access opportunities and collaborate across geographic and cultural boundaries" (Berners-Lee 2017). However, despite being a global common, the Web's wide penetration also makes it into a dominant standard. Through its ubiquity, the Web exerts pressure toward uptake, even if this uptake may harm the individual user. The alternative – refusing to be part of it – results in isolation. This phenomenon, which is described by David Grewal as *network power*, is common for networked standards (Grewal 2008, pp. 20–28). It makes the digital society into a hegemonic system from which – especially from the perspective of the Global South – there is no escape, despite the price users, communities, and even countries have to pay with their money or data, to become part of it.

When we observe the current structure of the digital society, we see that it is physically, economically, and socially extremely centralized and concentrated in the Global North, where to date the forbearers of many digital innovations reside. For

---

[1] See, e.g., https://www.un.org/development/desa/en/news/administration/internet-governance-2.html (Accessed 1 May 2021)

example, the "cloud" is concentrated in large datacenters in wealthy countries. The commercialization of ICTs, influenced by the said centralization, further puts a large chunk of any wealth gathered by innovations in the Global South into the accounts of Big Tech (Zuboff 2019, pp. 63–96). The unequal competition in terms of storage, connectivity, funding, and adoption hampers startup-driven innovation in the Global South.

While digital technologies such as mobile phones are becoming cheaper and more widespread in all corners of the world, control over what can be installed sits in the hands of prominent private tech firms. Governance and decision-making of technology are in the hands of the private tech firms and still, at best, bound by norms and regulations set in countries of the Global North. And these are just a few examples of technological coloniality.

Technological coloniality can be observed in many sectors of society. In Argentina, for example, a country that lacks technological autonomy, the digital market is dominated by transnational corporate tech firms. These parties are taking over roles and functions from the State, for example, in education. They are providing – through philanthropic gifts in the frame of so-called corporate social responsibility – digital services to higher education institutes in exchange for market penetration, tax savings, branding, and policy influencing. The commercial activities of the Big Tech companies are targeting youngsters in particular with media, music, video, entertainment, and fake news. This further leverages trends of privatization (Pini 2020, pp. 37–40).

With new forms of digital communication and online education, intensified in 2020 during the COVID-19 pandemic, there is growing evidence of algorithms being used for surveillance of access, production, and circulation of information, goods, and services in society. These scenarios are seen in many countries, Argentina included. "Free" Internet, provided in exchange for user data, is the business model in which personal data are exploited as raw material (Zuboff 2019, pp. 70–73). Data, knowledge, expertise, and high-performance infrastructures are kept and mined by an increasingly smaller number of transnational corporations, using highly advanced digital technologies for value extraction and profit. While interventions "free of costs" and "free Internet connectivity" are justified as societal benefits, the influence of the private tech sector in vital sectors of society reveals the corporate coloniality.

Coloniality can be observed in many instances of technology. For example, in Artificial Intelligence (AI) algorithms, which previously were held to be objective and value-free, discriminating biases have been discovered (Mohamed et al. 2020, pp. 659–663). There are various examples of discriminating AI, as a result of biases that are hidden in the underlying data: e.g. an algorithm that autonomously whitens black and Asian faces; an application, based on a face recognition algorithm, that opens the door of an office for white faces only, but fails to recognize black faces.[2] These trivial examples show biases embedded in apparently value-free technology, in which existing patterns are unconsciously replicated. These biases pop up unexpectedly in autonomous smart systems and may intentionally or unintentionally exacerbate inequalities. Artificial Intelligence is a technological domain that urgently needs to be innovatively decolonized.

2 Community-Oriented, Transdisciplinary Models and Inclusive Platforms as Alternative

At the brink of new technological breakthroughs, many scientists, aware of their responsibility, propose to bring together the brightest minds from various sectors and disciplines to discuss directions and propose solutions for the digital society (e.g., Berners-Lee 2019). The authors of this paper stress the importance to include, in these important platforms, also people from poor regions, e.g., in the Global South, and make their voices heard and their perspectives visible. To do so, we propose community-oriented research and collaborative technology development. While this can offshoot innovation in low-resource environments in unexpected ways, it can also be a source of inspiration for new forms of transdisciplinary knowledge production. We discuss a few examples.

In low-resource environments in Africa, many people do not have access to information which is relevant for their daily work. For example, smallholder farmers need local weather forecasts and data on actual rainfall and information on prices at local markets, on treatment of animal health, on the water quality in local wells, etc. However, information access is hampered not only due to a lack of Internet access: also cultural and social factors exist, for example, low literacy or language. These access barriers exist for the majority of rural communities in the Northern Region of Ghana.

In response to local needs, an exploratory design-science action-research project, dubbed Tibaŋsim, was carried out in Northern Ghana, to develop new modes of digital access and information sharing for rural communities. Tibaŋsim was deployed in five communities from the East Gonja District of the Savannah Region of Ghana. These are typically small communities with about 20 to 30 households. In this project, the Tibaŋsim information system was developed, built on local initiatives and adapted to the local conditions. Tibaŋsim provides farming-related information that is being collected, (re-)produced, and entered into the system by the community members themselves, so that it can be shared locally. It uses only technologies that are locally available: voice-based (GSM) mobile telephony and local community radio. The information is delivered to the users in their own local language(s) (Dittoh et al. 2021, pp. 1–23). Here, we see that it is not just about

---

[2] https://www.oneworld.nl/lezen/discriminatie/racisme/zwart-dan-rijdt-de-zelfrijdende-auto-jou-eerder-aan/ (Accessed: 1 May 2021)

connectivity or platform access as such: the collaborative work on relevant and adequate information content is at least as important.

Similar initiatives have been carried out in Mali in the period 2011–2021. At the request of the national Malian smallholder farmer organization AOPP,[3] a digital platform was developed to support their members – smallholder cereal seeds producers – in the seed trade. As soon as the first version of the Web-based seed trade platform was evaluated, it became clear that local requirements and contextual barriers had been overlooked by the technical developers (Vos et al. 2020, pp. 13–14). The system had to be adapted and re-designed in closer dialogue with its users. This second iteration resulted in a mobile, voice interface, spoken in the local language Bambara, as to be useful for farmers without literacy skills. This resulted in a complex set of requirements, as the platform should meet the requirements of the legacy non-digital local seed trade, the language and speech requirements, as well as the technical challenges to make it work in the absence of ubiquitous Internet connectivity.

In Sarawak, Malaysia, researchers and indigenous communities have worked together for over a decade, in search of sociotechnical solutions for local problems. One of the initiatives is eBario,[4] a project that aimed at connecting the unconnected remote village of Bario to the digital society. The initiative consisted of a transdisciplinary university-community partnership between one of Borneo ethnic minorities, the Kelabit community, and the Institute of Social Informatics and Technological Innovations of the University Malaysia Sarawak. The project brought many unexpected spin-offs for the indigenous community, who took joint knowledge creation as a new pathway. The joint efforts transformed the project into a living lab for innovations in healthcare, local cultural preservation, and agriculture. The eBario model has been replicated in six sites: Long Lamai and Ba'Kelalan sites in Sarawak, Pos Lenjang and Pos Sinderut sites in Pahang, and Pos Gob and Pos Bala sites in Kelantan. Among its achievements is the development of community-led, lifelong learning initiatives. Improved skills, incomes, and social communications were outcomes of the eBario project for participating communities. At a national level, the project has influenced policy-making for rural development. For academics, it brought new insights how to do ICT4D research that also seeks to improve the lives of marginalized and underserved communities (Harris et al. 2018, pp. 63–68). Projects such as these show that it is not just about universal connectivity per se: significant effort has to be spent on collaboratively shaping the societal impacts of connectivity as the key to reaping benefits of digitalization.

What we learn from the above initiatives in low-resource environments is that whereas mainstream computer science and Artificial Intelligence are only focusing on high-performance systems, high-end computing, networking, and big data, it is also scientifically challenging and societally relevant to investigate how to design

---

[3] https://aopp-mali.com/ (Accessed 1 May 2021)

[4] https://www.itu.int/osg/spuold/wsis-themes/ict_stories/themes/case_studies/e-bario.html (Accessed 1 May 2021)

small-scale solutions, decentralized systems, and green, energy-efficient technologies. For example, recent studies on small, inexpensive devices as the so-called "Kasadaka" platform[5] have demonstrated the potential of decentralized, inexpensive platforms, hosted locally on small hardware as inclusive platforms for local communities in Mali, Burkina Faso, and Ghana (Baart et al. 2019, pp. 202–219).

Another important point in this research is that of *contextualization*. For example, deployment of Artificial Intelligence generally requires high- performance infrastructures. Artificial Intelligence's most popular branch, Machine Learning, uses heavy computing and needs sustainable data storage to process and store large amounts of data. Such infrastructure is not available in many countries of the Global South. Another issue is related to user data which raises privacy and security issues and requires regulatory frameworks that unfortunately are still in their infancy in many African countries. Still, there are alternative forms of AI, e.g., knowledge-based reasoning systems, that will work better under low-resource circumstances and can run on decentralized local systems. Examples include knowledge engineering for indigenous knowledge, co-designed by local farmers and AI specialists, or expert systems of traditional African medicine co-developed by local and AI experts (Lô et al. 2017). These topics are currently being studied in field-based pilot research projects.

The above examples are real-world research projects with a modest reach. Despite their small size, these projects show the importance of transdisciplinarity, involving local communities, not as passive subjects, but as co-researchers and co-creators. This model is also applicable in academic education. Currently, mainstream curricula in Computer Science and Artificial Intelligence introduce students only to high-end domains of technological innovation. Not many educational programs are devoted to community-centered technology development in resource-constrained environments. Yet, the challenges of the global digital society are calling also for ICT professionals with the knowledge, skills, and responsibility to deal with these challenges. Collaborative technology development, reflection, and joint deliberation with respect for local agency and innovation can open new avenues toward responsible and societally oriented knowledge production and ethical technology development, striving for more equality and less coloniality in the (digital) society.

3   Conclusion

From the above discussions, it becomes clear that coloniality is a reality also in the digital society. Universal Internet connectivity does not necessarily equate to truly inclusive connectedness. According to African philosopher Achille Mbembe, we must realize that coloniality is more than academic discourses and representations (Mbembe 2001). It is a systemic problem, materialized in the real world and felt in

---

[5]https://www.kasadaka.com/ (Accessed 1 May 2021)

everyday life by many people. If we want to build a more human-centered, participatory, and democratic digital society – inclusive also for the most vulnerable communities – new ways of collaboration, innovation, and co-creation are needed. In this chapter, we have tried to present some directions on how this could be done.

References

Baart, A., Bon, A., De Boer, V., Dittoh, F., Tuijp, W. and Akkermans, H. (2019) "Affordable Voice Services to Bridge the Digital Divide – Presenting the Kasadaka Platform" in Escalona, M.J., Mayo, F.D., Majchrzak, T.A., Monfort, V. (eds) *Web Information Systems and Technologies,* LNBIP Book Series, Vol. 327, pp. 195-220. Berlin, Germany: Springer.

Berners-Lee, T. (2017) Three Challenges for the Web, *According to its Inventor.* [Online]. Available at: https://webfoundation.org/2017/03/web-turns-28-letter/ (Accessed: 1 May 2021).

Berners-Lee, T. (2019) The Web is under Threat. *Join us and Fight for it.* [Online] Available at: https://webfoundation.org/2018/03/web-birthday-29/ (Accessed: 1 May 2021).

Dittoh, F., Akkermans, H. De Boer, V. Bon, A. Tuyp, W. and Baart, A. (2021) "Tibaŋsim: Information Access for Low-Resource Environments" in Yang, X.S., Sherratt, S., Dey, N., Joshi, A. (eds) *Proceedings of the Sixth International Congress on Information and Communication Technology*: ICICT 2021, London, UK, Vol. 1, Singapore: Springer. Available at: https://w4ra.org/wp-content/uploads/2014/02/ICICT_2021_paper_289.pdf (Accessed: 1 May 2021).

Lee, E.A. (2020) The Coevolution: The Entwined Futures of Humans and Machines. Cambridge MA, USA: MIT Press.

Grewal, D. S. (2008) Network Power: The Social Dynamics of Globalization. New Haven, USA & London, UK: Yale University Press.

Harris, R., Ramaiyer, N.A.N.K. and Tarawe, J. (2018) "The eBario Story: ICTs for Rural Development" In *International Conference on ICT for Rural Development (ICICTRuDev)* pp. 63-68, IEEE.

Mignolo, W.D. and Walsh, C.E. (2018) On Decoloniality: Concepts, Analytics, Praxis. Durham, NC, USA: Duke University Press.

Mendoza, B. (2021) "Decolonial Theories in Comparison" in Shih S., Tsai, L. (eds) *Indigenous Knowledge in Taiwan and Beyond.* Sinophone and Taiwan Studies, Vol .1, pp. 249-271, Singapore: Springer.

Lô, G., de Boer, V., Schlobach, S. and Diallo, G. (2017) "Linking African Traditional Medicine Knowledge". *Semantic Web Applications and Tools for Healthcare and Life Sciences* (SWAT4LS), Rome Italy. [Online] Available at: https://hal-archives-ouvertes.fr/hal-01804941/document (Accessed 1 May 2021).

Mohamed, S., Png, M.T. and Isaac, W. (2020) "Decolonial AI: Decolonial Theory as Sociotechnical Foresight in Artificial Intelligence". Philosophy & Technology, Vol. 33 No. 4, pp. 659-684.

Mbembe, A. (2001) *On the Postcolony.* Studies on the History of Society and Culture, Vol. 41, Los Angeles, USA: University of California Press.

Pini, M.E. (2020) "Digital Inequality in Education in Argentina". In *Proceedings of the 12th ACM Conference on Web Science* (WebSci '20 Companion), July 6–10, 2020, Southampton, UK, pp. 37-40, New York, NY, USA: ACM. Available at: https://doi.org/10.1145/3394332.3402827.

Quijano, A. (2016) "Bien Vivir – Between Development and the De/Coloniality of Power". *Alternautas (Re) Searching Development: The Abya Yala Chapter* 3(1) pp. 10-23. [Online] Available at: http://www.alternautas.net/blog/2016/1/20/bien-vivir-between-development-and-the-decoloniality-of-power1 (Accessed 1 May 2021).

Vos, S., Schaefers, H., Lago, P. and Bon, A. (2020) "Sustainability and Ethics by Design in the Development of Digital Platforms for Low-Resource Environments". [Online]. *Amsterdam Sustainability Institute Integrative Project Technical Report.* pp. 1-43, Vrije Universiteit Amsterdam. Available at: https://w4ra.org/wp-content/uploads/2021/01/ICT4FoodSec.pdf (Accessed 1 May 2021).

Zuboff, S. (2019) The Age of Surveillance Capitalism: The Fight for a Human Future at the New Frontier of Power. London, UK: Profile Books.

Part III
Ethics and Philosophy of Technology

Digital Humanism and the Limits of Artificial Intelligence

Julian Nida-Rümelin

Abstract This chapter is programmatic in style and content. It describes some patterns and one central argument of that, what I take as the view of digital humanism and which we exposed in our book (Nida-Rümelin and Weidenfeld 2018). The central argument regards the critique of strong and weak AI. This chapter does not discuss the logical and metaphysical aspects of digital humanism that I take to be part of the broader context of the theory of reason (Nida-Rümelin 2020, Chaps. VI and VII).

I

The expression "Artificial Intelligence" (AI) is multifaceted and is used with different meanings. In the broadest and least problematic sense, AI denotes everything from computer-controlled processes, the calculation of functions, the solution of differential equations, logistical optimization, and robot control to "self-learning" systems, translation software, etc. The most problematic and radical conception of AI says that there is no categorical difference between computer-controlled processes and human thought processes. This position is often referred to as "strong AI." "Weak AI" then merely is the thesis that all thought and decision processes could in principle be simulated by computers. In other words, the difference between strong and weak AI is the difference between identification and simulation. From this perspective, strong AI is a program of disillusionment: What appears to us to be a characteristically human property is nothing but that which can be realized as a computer program. Digital humanism takes the opposite side.

J. Nida-Rümelin (✉)
Ludwig Maximilians Universität Munich, Munich, Germany
e-mail: julian.nida-ruemelin@lrz.uni-muenchen.de

II

The analytic philosopher John Searle (1980) has devised a famous thought experiment. Searle asks us to imagine yourself being a monolingual English speaker "locked in a room and given a large batch of Chinese writing" plus "a second batch of Chinese script" and "a set of rules" in English "for correlating the second batch with the first batch." The rules "correlate one set of formal symbols with another set of formal symbols": "formal" (or "syntactic") meaning you "can identify the symbols entirely by their shapes." A third batch of Chinese symbols and more instructions in English enable you "to correlate elements of this third batch with elements of the first two batches" and instruct you, thereby, "to give back certain sorts of Chinese symbols with certain sorts of shapes in response." *Those giving you the symbols* "call the first batch 'a script'" [a data structure with natural language processing applications], "they call the second batch 'a story,' and they call the third batch 'questions'"; the symbols you give back "they call. .. 'answers to the questions'"; "the set of rules in English. .. they call 'the program'": *you yourself* know none of this. Nevertheless, you "get so good at following the instructions" that "from the point of view of someone outside the room," your responses are "absolutely indistinguishable from those of Chinese speakers." Just by looking at your answers, nobody can tell you don't speak a word of Chinese. Outside in front of the slot, there is a native speaker of Chinese, who, having formulated the story and the questions and having received the answers, concludes that somebody must to be present in the room who also speaks Chinese.

The crucial element missing here is apparent: It is the understanding of the Chinese language. Even if a system—in this case the Chinese Room—is functionally equivalent to somebody who understands Chinese, the system does not yet itself understand Chinese. Understanding and speaking Chinese requires various kinds of knowledge. A person who speaks Chinese refers with specific terms to the corresponding objects. With specific utterances, she pursues certain—corresponding—aims. On the basis of what she has heard (in Chinese), she forms certain expectations, etc. The Chinese Room has none of these characteristics. It does not have any intentions; it has no expectations that prove that it speaks and understands Chinese. In other words, the Chinese Room simulates an understanding of Chinese without itself possessing a command of the Chinese language.

Years later, Searle (1990) radicalized this argument in connecting it with philosophical realism (Nida-Rümelin 2018), that is, the thesis that there is a world that exists regardless of whether it is observed or not. Signs only have a meaning for us, the sign users and sign interpreters. We ascribe meaning to certain letters or symbols by communicating, by agreeing that these letters or symbols stand for something. They have no meaning without these conventions. It is misleading to conceive the computer as a character-processing, or syntactic, machine that follows certain logical or grammatical rules. The computer is comprised of various elements that can be described by physics, and the computational processes are a sequence of electrodynamic and electrostatic states. To these states, signs are then ascribed, to which we attribute certain interpretations and rules. The physical processes in the computer have no syntax, they do not "know" any logical or grammatical rules, and they are not even strings of characters. The syntactical interpretation is observer-relative. As syntactic structures are observer-relative, the world is not a computer. This argument is radical, simple, and accurate. It rests on a realist philosophy and a mechanistic interpretation of computers. Computers are that which they are materially: objects that can be completely described and explained using the methods of physics. Syntax is not a part of physics; physics describes no signs, no grammatical rules, no logical conclusions, and no algorithms. The computer simulates thought processes without thinking itself. Mental properties cannot be defined by behavioral characteristics. The model of the algorithmic machine, of mechanism, is unsuitable as a paradigm both for the physical world and as a paradigm for human thinking.

A realist conception is far more plausible than a behaviorist conception regarding mental states (Block 1981). Pains characterize a specific type of feelings that are unpleasant and that we usually seek to avoid. At the dentist, we make an effort to suppress any movement so that we do not interfere with the treatment, but by no means does this mean that we have no pain. Even the imaginary super-Spartan, who does not flinch even under severe pain, can have pain. It is simply absurd to equate "having pain" with certain behavioral patterns.

III

It can be shown that logical and mathematical proofs to a large extent cannot be based on algorithms, as students of formal logics learn early on in their study. Already the calculi of first-order predicate logic do not allow for algorithmic proof writing. The fundamental reason for this phenomenon, that more complex logical systems than propositional logic are not algorithmic in this sense, is Kurt Gödel's incompleteness theorem (Gödel 1931), the probably most important theorem of formal logic and meta-mathematics. This theorem shows that insight and intelligence in general cannot be grasped adequately within a machine paradigm (Lucas 1961). One can interpret Gödels theorem as the proof that the human mind does not work like an algorithm. Possibly even consciousness in general is based on incompleteness as Roger Penrose (1989) argues, but I remain up to now agnostic about this question, being however convinced that neither the world nor human beings function like a machine.

If humans were to act just as deterministically as Turing machines (Turing 1950), then genuine innovation itself would not be imaginable. If it was in principle possible to foresee what we do and believe in the future, genuine innovations would not exist. Disruptive innovations in knowledge and technology require that future knowledge and technology is not part of old knowledge and technology. The assumption of an all-comprising determinism is incompatible with true innovation (Popper 1951, 1972). It is more plausible to assume that the thesis of weak AI, the thesis that all human deliberation can be simulated by software systems, is wrong, than to assume that there is no genuine innovation.

IV

Digital humanism advocates the employment of digital technologies in order to improve human living conditions and preserve ecological systems, also out of concern for the vital interests of future generations. At the same time, however, it vehemently opposes a supposedly autarchic technological development of digital transformation. It opposes the self-depreciation of human competence in deciding and acting in the form of strong and weak AI; it opposes the subsumption of human judgment and agency under the paradigm of a machine that generates determined outputs from given inputs.

The utopia of digital humanism demands a consistent departure from the paradigm of the machine. Neither nature as a whole nor humans should be conceived of as machines. The world is not a clock, and humans are not *automata*. Machines can expand, even potentiate, the scope of human agency and creative power. They can be used for the good and to the detriment of the development of humanity, but they cannot replace the human responsibility of individual agents and the cultural and social responsibility of human societies. Paradoxically, the responsibility of individuals and groups is broadened by machine technology and digital technologies. The expanded possibilities of interaction enabled through digital technologies and the development of communicative and interactive networks rather present new challenges for the ethos of responsibility, which the rational human being cannot evade by delegating responsibility to autonomous systems, be they robots or self-learning software systems.

Digital humanism retains the human conditions of responsible practice. It does not commit a category mistake. It does not ascribe mental properties based on a simulation of human behavior. Rather, it sharpens the criteria of human responsibility in the face of the availability of digital technologies, calls for an expansion of the ascription of responsibility to communication and interaction mediated by digital technologies, and does not allow the actual agents (and that is us humans) to duck away and pass responsibility on to a supposed autonomy of digital machines. Digital humanism is directed at strengthening human responsibility, at realizing the potentials of digitalization that relieve the burden of unnecessary knowledge and calculations in order to give people the possibility to concentrate on what is essential and contribute to a more humane and just future for humanity.

References

Block, Ned (1981), Psychologism and Behaviorism, The Philosophical Review 90 (1): 5–43.
Gödel, Kurt (1931), Über formal unentscheidbare Sätze der Principia Mathematica und verwandter Systeme I, Monatshefte für Mathematik und Physik 38: 173–198.
Lucas, John R. (1961), On Minds, Machines and Gödel, *Philosophy* 36: 112–127.
Nida-Rümelin, Julian (2018), Unaufgeregter Realismus. *Eine philosophische Streitschrift*, Paderborn: mentis.
Nida-Rümelin, Julian (2020), Eine Theorie praktischer Vernunft, Berlin/Boston: De Gruyter.
Nida-Rümelin, Julian, Weidenfeld, Nathalie (2018), *Digitaler Humanismus. Eine Ethik für das Zeitalter der Künstlichen Intelligenz*, München: Piper (italian translation: Milano: Franco Angeli 2019; korean translation: Pusan National University Press 2020).
Penrose, Roger (1989), The Emperor's New Mind: Concerning Computers, Minds, and the Laws of Physics, Oxford University Press.
Popper, Karl (1951), Indeterminism in Quantum Physics and Classical Physics, British Journal of Philosophy of Science 1: 179–188.
Popper, Karl (1972), Objective Knowledge, Oxford University Press.
Searle, John (1980), "Minds, Brains and Programs", *Behavioral and Brain Sciences* 3 (3): 417–457.
Searle, John (1990), "Is the Brain a Digital Computer?", Proceedings and Addresses of the American Philosophical Association, 64 (3): 21–37.
Turing, Alain (1950): Computing Machinery and Intelligence, Mind 59: 433–460.

Explorative Experiments and Digital Humanism: Adding an Epistemic Dimension to the Ethical Debate

Viola Schiaffonati

Abstract The rise of Digital Humanism calls for shaping digital technologies in accordance with human values and needs. I argue that to achieve this goal, an epistemic and methodological dimension should be added to the ethical reflections developed in the last years. In particular, I propose the framework of explorative experimentation in computer science and engineering to set an agenda for the reflection on the ethical issues of digital technologies that seriously considers their peculiarities from an epistemic point of view. As the traditional epistemic categories of the natural sciences cannot be directly adopted by computer science and engineering, the traditional moral principles guiding experimentation in the natural sciences should be reconsidered in the case of digital technologies where uncertainty about their impacts and risks is very high.

1 Introduction

The rise of Digital Humanism calls for shaping digital technologies in accordance with human values and needs to possibly solve the critical issues of current technological development. Within this framework, ethics plays an increasing role at both a descriptive level and normative one, and, accordingly, several important results have been achieved in the last years. On the one hand, approaches such as the Value Sensitive Design have shifted the attention to the idea of *active responsibility*, that is, the design of technology to incorporate positive values (van den Hoven 2007). On the other hand, several regulatory frameworks have been proposed to address the ethical issues related to digital technologies, such as AI, and their adoption within our society.

Notwithstanding the importance of these initiatives, I argue that a further dimension should be added to this debate. This dimension concerns the analysis of the disciplinary and methodological status of computer science and engineering to better

V. Schiaffonati (✉)
Politecnico di Milano, Milan, Italy
e-mail: viola.schiaffonati@polimi.it

H. Werthner et al. (eds.), *Perspectives on Digital Humanism*,
https://doi.org/10.1007/978-3-030-86144-5_11

77

understand the radical paradigm shift promoted by digital technologies. Rather than considering this dimension as alternative to the other ones, I claim that it should be integrated with them to address the current challenges of digital technologies in a more comprehensive way. In this chapter, I focus in particular on the nature and role of experiments in AI and autonomous robotics. The main result of adding this further dimension to the current analysis is to set an agenda for the reflection on the ethical issues of digital technologies that seriously considers their peculiarities from a disciplinary and a methodological point of view. Constructing on some of my previous works, I argue that the traditional epistemic categories of the natural sciences cannot be directly adopted by computer science and engineering as an artificial discipline. Accordingly, the traditional moral principles guiding experimentation in the natural sciences should be reconsidered in the case of digital technologies, where uncertainty about their impacts and risks is very high.

This chapter is organized as follows. Section 2 discusses the nature and role of experiments in computer science and engineering and how experiments are perceived as ways to increase the scientific maturity of the field. Section 3 presents the novel notion of explorative experimentation emerging from the analysis of the practice of AI and autonomous robotics. Section 4 connects epistemic uncertainty, typical of explorative experiments, to the design of ethical frameworks based on an incremental approach. Finally, Sect. 5 concludes the chapter by stressing how explorative experiments can impact on the current shaping of Digital Humanism.

2   Experimental Method and Computing

In the last years, the debate on the nature and role of experiments in computer science and engineering has emerged as one of the ways to stress its scientific status: to adopt the same experimental standards of the natural sciences can make computer science and engineering more mature and credible.

AI and autonomous robotics make no exception. AI, for example, is facing a reproducibility crisis in which the importance of reproducibility is taken for granted: the specificity of reproducibility in AI is not investigated, and, in the end, only practical benefits are evidenced (Gundersen et al. 2018). Autonomous robotics presents two different tendencies (Amigoni et al. 2014). On the one hand, the traditional principles of experimental method (reproducibility, repeatability, generalization, etc.) are seen as golden standards to which the research practice should conform. For example, public distribution of code is promoted to achieve reproducibility. On the other hand, rigorous approaches to experimentation are not yet part of current practices. For example, the use of settings that can be applied to different environments is limited, jeopardizing the possibility of generalizing experimental results.

Only few exceptions have stressed the peculiarity of experimentation in computer science and engineering and emphasized that the term experiment can be used in different ways (Tedre 2015). Moreover, the question whether it does make sense to apply the same standards of the natural sciences to the artificial ones has been seldom asked. The idea that computer science and engineering is an experimental science of a very special type has been advanced by Allen Newell and Herbert Simon already in the 1970s (Newell and Simon 1976). Even if the invitation to see each new machine as an experiment has remained largely unattended, some exceptions exist: they point out that experimentation is more multifaceted than usually depicted in computer science and engineering.

Two elements are particularly important. First, many experiments have the goal of testing technical artifacts rather than theories. *Technical artifacts* are physical objects with a technical function and use plan designed by humans in order to fulfill some practical functions (Veermas et al. 2011). Second, in several cases, experimenters are designers, thus losing the independence of the experimenter prescribed in the classical experimental protocol. This is why I have proposed the notion of *explorative experimentation* to give reason of a part of the experimental practice in computing that cannot be subsumed under the traditional categories of the epistemic and controlled experimentation typical of the natural sciences (Schiaffonati 2020).

3 A Different Notion of Experimentation: Explorative Experiments

Explorative experiments are a technological form of experimentation devoted to test technical artifacts. They can be seen in continuity with the tradition of the so-called directly action-guiding experiments, that is, those experiments devoted to action and contraposed to traditional *epistemic experiments* devoted to knowledge (Hansson 2015). For example, a systematic test on an autonomous robot employed to assist an elderly person in her home is a technological form of experimentation, where the outcome looked for is the proper interaction of the robot with the person and the intervention is the careful tuning of the abilities of the robot to achieve the goal.

Moreover, explorative experiments have a normative component that epistemic ones do not possess. They are carried out to check whether the technical artifacts meet the desired specifications via their technological production. The normative element consists in determining how much the tested technical artifact conforms to its design. A robot system, for example, can be evaluated as better or worse with respect to a given function that works as a reference model. On the contrary, a natural phenomenon (which is usually what is investigated in an experiment in the natural sciences), such as an electron, cannot be good or bad: the electron in the experiment is evaluated without any reference to its supposed technical function, hence without any normative constraint with respect to its correct functioning.

To summarize, explorative experiments are devoted to test technical artifacts without the control boundaries typical of an epistemic controlled experiment. Their goal is to investigate the possibilities and limits of the technical artifact and its interaction with the surrounding environment. The design of this investigation is not conducted on the basis of a well-formed theory or a systematic theoretical background. Rather, the initial hypotheses cannot always be formulated in a clear way, and the type of knowledge which is the goal of this experimentation is oriented to evaluate the performance of the technical artifact with respect to its technical function. The experimenter is often the same designer of the artifact, and, thus, her independence from the experimental context, as in the traditional epistemic experimentation, is not guaranteed. In conclusion, explorative experiments are not devoted to reject or accept a general theory, but to probe (iteratively) the possibilities and limits of the intervention. This makes them similar to some methodological reflections developed in the field of design science research, where the iterative and evolutionary nature of design improvements through exploration is emphasized (Gil and Hevner 2013).

4 From Epistemic Uncertainty to Ethical Incrementalism

The explorative experiment framework highlights how uncertainty plays an essential role at a theoretical level and how this has an impact on the experimental procedures that must renounce to a part of the experimental control traditionally associated to epistemic experiments. When considering AI and autonomous robotics, uncertainty concerns both the behavior of the complex systems themselves and their interactions with humans and complex environments. For this reason, they can be labelled *experimental technologies*: the operational experience relative to their effective behavior is limited, and, therefore, the attempts to precisely assess their societal risks and benefits are uncertain: this means that their impact on humans and societies is mostly unknown and difficult to predict (van de Poel 2016). To acknowledge uncertainty in the development of experimental technologies means to recognize that unexpected events can always occur and that a different approach is required to deal with their development and management. This approach is a form of *incrementalism*, where experimental technologies are gradually introduced into society to constantly monitor the societal effects that emerge and iteratively improve their design accordingly. In other words, the epistemic uncertainty emerging from an epistemological perspective in the case of explorative experimentation can be translated into a form of incrementalism from an ethical perspective. Explorative experiments, devoted to acquiring knowledge on the behavior of these experimental systems in the real world, are therefore crucial to address the ethical issues related to the impact on such systems on society.

Some ethical frameworks have already proposed to deal with these experimental technologies. For example, van de Poel (2016) incorporates the traditional principles of bioethics (beneficence, non-maleficence, respect for autonomy, respect for justice) and declines them in an incremental ethical framework. I have argued elsewhere that this a promising starting point in particular when integrated with explorative experiments (Amigoni and Schiaffonati 2018). This means that to concretely minimize the risks associated to experimental technologies, the first step is to understand what it really accounts to experiment on technical artifacts. The framework of explorative experiments has thus an impact not only at the methodological level but also at the ethical one, where the traditional moral categories need to be revised to deal with experimental technologies.

5 Conclusion

In this chapter, I have suggested that to address some of the issues connected to digital technologies, the development of appropriate techniques is not enough. Rather, I have shown that some problems have to be addressed with methods having a philosophical nature.

To conclude, I emphasize how the framework of explorative experimentation is connected to the larger issue of the societal impact of digital technologies. The problem of how this approach can be better adopted in practice remains open. Yet I argue that a shift in the conceptualization has, at least, two important roles. The first one concerns the influence of novel epistemic categories, such as explorative experiments, on ethical ones, as I have discussed in Sect. 4. The second level regards the development of the disciplines of the artificial, to which computer science and engineering belong, by starting from methodological reflections. This is not only a disciplinary issue, but has an impact on how humans, digital technologies, and their interactions are conceptualized in the current discussion on Digital Humanism. If one of the goals of Digital Humanism is to "shape technologies in accordance with human values and needs, instead of allowing technologies to shape humans," it is essential to recognize the centrality of technical artifacts and sociotechnical systems in the disciplines of the artificial. Sociotechnical systems are composed of physical objects, people, organizations, institutions, conditions, and rules. They, thus, have a hybrid character as they consist of components which belong in many different "worlds": not only those requiring a physical description but also those requiring a social one (Veermas et al. 2011). So far, the components requiring a physical description have been addressed by scientific and engineering disciplines. Now it is time to consider all the components requiring a social description, like the ones promoted by the humanities and the social sciences, and to develop, accordingly, the new field of the artificial disciplines which should include and integrate both in a creative way.

References

Amigoni, F., Schiaffonati, V., Verdicchio, M. (2014) 'Good Experimental Methodologies for Autonomous Robotics: From Theory to Practice', in F. Amigoni, V. Schiaffonati (eds.), Methods and Experimental Techniques in Computer Engineering, SpringerBriefs in Applied Sciences and Technology, Springer, pp. 37-53.

Amigoni, F. and Schiaffonati, V. (2018) 'Ethics for Robots as Experimental Technologies: Pairing Anticipation with Exploration to Evaluate the Social Impact of Robotics', in *IEEE Robotics and Automation Magazine*, 25, n. 1, pp. 30-36.

Gil, T. G. and Hevner, A. N. (2013) 'A Fitness-Utility Model for Design Science Research', in ACM Transactions on Management Information Systems, 4 (2): 5-24.

Gundersen, O. E., Aha, D., Gil, Y. (2018) 'On Reproducible AI: Towards Reproducible Research, Open Science, and Digital Scholarship in AI Publications', in *AI Magazine*, 39, n. 3, pp. 56-68.

Hansson, S.O. (2015) 'Experiments before Science? – What Science Learned from Technological Experiments', in Sven Ove Hansson (ed.) *The Role of Technology in Science*, Springer.

Newell, A. and Simon, H. (1976) 'Computer Science as Empirical Inquiry: Symbols and Search', Communications of the ACM 19 (3): 113-126.

Schiaffonati, V. (2020). Computer, robot ed esperimenti, Milano: Meltemi.

Tedre, M. (2015) The Science of Computing. Boca Raton: CRC Press, Taylor & Francis Group.

van de Poel, I. (2016) 'An Ethical Framework for Evaluating Experimental Technology', in Science and Engineering Ethics, 22, pp. 667-686.

van den Hoven J. (2007) 'ICT and Value Sensitive Design', In Goujon P., Lavelle S., Duquenoy P., Kimppa K., Laurent V. (eds.) *The Information Society: Innovation, Legitimacy, Ethics and Democracy In honor of Professor Jacques Berleur s.j*, IFIP International Federation for Information Processing, vol 233, Boston: Springer.

Veermas, P., Kroes, P., van de Poel, I., Franssen, M., Houkes, W. (2011) A Philosophy of Technology: From Technical Artefacts to Sociotechnical Systems, Morgan & Claypool Publishers.

Digital Humanism and Global Issues in Artificial Intelligence Ethics

Guglielmo Tamburrini

Abstract In the fight against pandemics and climate crisis, the zero hunger challenge, the preservation of international peace and stability, and the protection of democratic participation in political decision-making, AI has increasing – and often double-edged – roles to play in connection with ethical issues having a genuinely global dimension. The governance of AI ambivalence in these contexts looms large on both the AI ethics and digital humanism agendas.

1 Introduction

Global ethical issues concern humankind as a whole and each member of the human species irrespective of her or his position, functions, and origin. Prominent issues of this sort include the fight against pandemics and climate crisis, the zero hunger challenge, the preservation of international peace and stability, and the protection of democracy and citizen participation in political decision-making. What role is AI playing – with its increasingly pervasive technologies and systems – and will be likely to play in connection with these global ethical challenges?

The COVID-19 pandemic has raised distinctive challenges of human health and well-being protection across the planet, which are inextricably related to worldwide issues of economic resilience and protection of education, work, and social life participation rights. Artificial Intelligence (AI) has the potential to become a major technological tool to meet pandemic outbursts and the attending ethical issues. Indeed, infection spreading data and machine learning (ML) technologies pave the way to computational models for predicting diffusion patterns, identifying and assessing the effectiveness of pharmacological, social, and environmental measures, up to and including the monitoring of wildlife ecological niches, whose preservation is so important to restrain frequent contacts with wild animal species and related virus spillovers. Similarly, AI affords technological tools to optimize food

G. Tamburrini (✉)
Università di Napoli Federico II, Naples, Italy
e-mail: guglielmo.tamburrini@unina.it

production and distribution so as to fight famines and move toward the zero hunger goal in the UN sustainable development agenda.

Failures to use effective AI technologies to fight pandemics and world hunger may qualify as morally significant omissions. Along with these omissions, another source of moral fault may emerge from the ethically *ambivalent* roles that AI is actively assuming in the context of other global challenges. On the one hand, AI models may contribute to identify energy consumption patterns and corresponding climate warming mitigation measures. On the other hand, AI model training and related big data management produce a considerable carbon footprint. Similarly, AI military applications may improve Communications, Command, and Control (C3) networks and enhance both precision and effectiveness of weapons systems, leading to a reduction of military and civilian victims in warfare situations. And yet, the ongoing AI arms race may increase the tempo of conflicts beyond meaningful human control and lower the threshold to start conflicts, thereby threatening international peace and stability. Just as importantly, AI systems may help one in retrieving the diversified political information which is needed to exercise responsible democratic citizenship. However, in both authoritarian and democratic countries, AI systems have been already used to curtail freedom and participation in political decision-making.

As exemplary cases of AI playing ambivalent roles in global ethical issues, I will focus here on the climate crisis and the preservation of global peace and international stability. Universal human values and needs that are prized by digital humanism play a crucial role in the governance of such AI ambivalence.

2 Artificial Intelligence Ethics and the Climate Crisis

AI models are well-suited to identify and monitor energy consumption patterns, in addition to suggest policy measures for curbing carbon emissions in transportation, energy, and other production sectors characterized by high carbon footprints. AI potential contribution to climate warming mitigation actions is extensively illustrated by the Climate Change AI Group (Rolnick et al. 2019) and advocated in multiple current initiatives of AI research and industry (https://aiforgood.itu.int). AI is presented there as a new technological opportunity to promote both intergenerational and intragenerational justice and to enact human responsibilities toward other living entities. However, exactly the same ethical values and responsibilities impel AI communities to look closely into the backyard of their own carbon footprint. The more optimistic forecasts suggest that the carbon footprint of the entire digital technology sector, including AI, will remain stable between now and 2050 (Blair 2020). But even this optimistic outlook is no reason for inaction. Indeed, if other production sectors will reduce their carbon footprint in accordance with the Paris Agreement goals, the proportion of global carbon emissions taking their origin in the ICT sector will considerably increase over the same temporal interval.

Within the widely differentiated ICT sector, extensive discussion is under way about the energy consumption of some non-AI software, like blockchain and other cryptocurrency software, which are estimated to consume amounts of energy exceeding the energy need of countries like Ukraine or Sweden (https://cbeci.org/cbeci/comparisons/). In contrast with this, it is still unclear which fraction of the ICT sector energy consumption can be specifically attributed to AI in general or to machine learning or other prominent research and commercial subfields in particular. Available data are mostly anecdotal. It was estimated that GPT-2 and GPT-3 – successful natural language processing (NLP) models for written text production developed by ML techniques – were trained by means of huge amounts of textual data and gave rise to a carbon footprint comparable to that of five average cars throughout their lifecycle (Strubell et al. 2019). More systematic assessment efforts are clearly needed.

Considering the increasingly pervasive impact of AI technologies, the *White Paper on AI* released in 2020 by the European Commission recommends addressing the carbon footprint of AI systems across their lifecycle and supply chain: "Given the increasing importance of AI, the environmental impact of AI systems needs to be duly considered throughout their lifecycle and across the entire supply chain, e.g., as regards resource usage for the training of algorithms and the storage of data" (EU 2020, p. 2). However, one should carefully note that developing suitable metrics and models for estimating the AI carbon footprint at large is a challenging and elusive problem. To begin with, it is difficult to precisely circumscribe AI within the broader ICT sector. Moreover, a sufficiently realistic assessment requires one to consider wider interaction layers between AI technologies and society, including AI-induced changes in work, leisure, and consumption patterns. These wider interaction layers have proven difficult to encompass and measure in the case of various other technologies and systems.

Without belittling the importance and the difficulty of achieving a sufficiently realistic evaluation, what is already known about the lifecycle of both exemplary AI systems like GPT-2 and GPT-3 and the supply chain of big data for ML suffices to spur a set of interrelated policy questions. Should one set quantitative limits to energy consumption for AI model training? How are AI carbon quotas, if any, to be identified at national and international levels? How to distribute equitable shares of AI limited resources to business, research, and public administration? Who should be in charge of deciding which data for AI training to collect, preserve, and eventually get rid of for the sake of environmental protection? (Lucivero 2019). Only by addressing these issues of environmental justice and sustainability can AI be made fully compatible with the permanence on our planet of human life and the unique moral agency that comes with it, grounding human dignity and the attending responsibilities that our species has toward all living entities (Jonas 1979).

3   Ethics and the Artificial Intelligence Arms Race

The protection of both human life and dignity has been playing a crucial role in the ethical and legal debate about autonomous weapons systems (AWS), that is, weapons systems that are capable of selecting and attacking military objectives without requiring any human intervention after their activation. The wide spectrum of positions emerging in this debate has invariably acknowledged as a serious possibility the occurrence of AWS suppressing human lives in violation of International Humanitarian Law (IHL) (Amoroso and Tamburrini 2020). Indeed, AI perceptual systems, developed by machine learning and paving the way to more advanced AWS, were found by adversarial testing to incur into unexpected and counter-intuitive errors that human operators would easily detect and avoid. Notable in the AWS debate context is the case of a school bus taken for an ostrich (Szegedy et al. 2014). Since properly used school buses and their passengers are protected by IHL distinction and proportionality principles, the example naturally suggests the following question: Who will be held responsible for unexpected and difficult to predict AWS acts that one would regard as war crimes, had they been committed by a human being?

The use of AWS has been additionally claimed to entail a violation of human dignity (Amoroso and Tamburrini 2020, p. 5). Robert Sparrow aptly summarized this view, pointing out that the decision to take another person's life must be compatible with the acknowledgement of the personhood of those with whom we interact in warfare. Therefore, "when AWS decide to launch an attack, the relevant interpersonal relationship is missing," and the human dignity of the potential victims is not recognized. "Indeed, in some fundamental sense there is no one who decide whether the target of the attack should live or die" (Sparrow 2016, pp. 106–7).

These various concerns about IHL and human dignity respect have been upheld since 2013 by the international Campaign to Stop Killer Robots in advocacy for a ban on lethal AWS. The Campaign has also extensively warned that AWS may raise special threats to international peace. The latter is a fundamental precondition for the flourishing of human life that any sensible construal of humanism as a doctrine and movement – including digital humanism – is bound to recognize as a highly prized value. AWS threaten peace by making wars easier to wage on account of reduced numbers of involved soldiers, by laying conditions for unpredictable runaway interactions between AWS on the battlefield, and by accelerating the pace of war beyond human cognitive and sensory-motor abilities.

AI may bring about threats to international peace and stability in the new cyberwarfare domain too. Indeed, AI learning systems are expected to become increasingly central there, not only for their potential to expand cyberdefence toolsets but also to launch more efficient cyberattacks (Christen et al. 2020, p. 4). Cyberattacks aimed at nuclear weapons command and control networks, at the hacking of nuclear weapons activation systems, or at generating false nuclear attack warnings raise special concerns. Accordingly, the confluence of AI cyberweapons with nuclear weapons intensifies that distinctive threat to the permanence on our planet of human life and moral agency that physicists and other scientists have been publicly denouncing at least since the Russell-Einstein Manifesto in 1955.

From the development of AWS to AI systems for discovering software vulnerabilities and waging cyberconflicts, an AI arms race is well under its way. The weaponization of AI should be internationally regulated and the AI arms race properly bridled. Digital humanism, with its analyses and policies inspired by universal ethical values and the protection of human dignity, has a central role to play in this formidable endeavor.

4 Concluding Remarks

The AI ethics agenda has been mostly concerned with ethical issues arising in specific AI application domains. Familiar cases are issues arising in connection with loans, careers and job hiring automatic decisions, insurance premium evaluation, or parole-granting tribunal judgments. Selectively affecting designated groups of stakeholders, these may aptly be called *local* ethical issues. Here, the focus has been placed instead on AI ethics issues that are *global*, insofar as they impact on humankind and all members of the human species as such. The climate crisis and the AI arms race have been used as exemplary cases to illustrate both the difference between local and global ethical issues and the need for a proper governance of AI ethically ambivalent roles. Last but not least, it has been argued that the ethical governance of this ambivalence makes crucial appeal to universal human values that any doctrine or movement deserving the name of digital humanism must endorse and support in the context of the digital revolution.

References

Amoroso D., Tamburrini G. (2020) 'Autonomous Weapons Systems and Meaningful Human Control: Ethical and Legal Issues', Current Robotics Reports 1(7), pp. 187–194. https://doi.org/10.1007/s43154-020-00024-3

Blair G. S., (2020) 'A tale of two cities: reflections on digital technology and the natural environment', Patterns 1(5). https://www.cell.com/patterns/fulltext/S2666-3899(20)30088-X

Christen M., Gordijn B., Loi M. (eds) (2020). The Ethics of Cybersecurity. Cham: Springer. https://link.springer.com/book/10.1007%2F978-3-030-29053-5

European Commission (2020). *White paper on AI. A European approach to excellence and trust*, Bruxelles, 19 February 2020. https://ec.europa.eu/info/sites/info/files/commission-white-paper-artificial-intelligence-feb2020_en

Jonas H. (1979). Das Prinzip Verantwortung. *Versuch einer Ethik für die technologische Zivilisation*. Frankfurt am Main: Insel-Verlag.

Lucivero, F. (2019) 'Big data, big waste? A reflection on the environmental sustainability of big data initiatives', Science and Engineering Ethics, 26, pp. 1009–30. https://doi.org/10.1007/s11948-019-00171-7

Rolnick D. et al. (2019) 'Tackling Climate Change with Machine Learning', *arxiv.org.1906.05433*. https://arxiv.org/abs/1906.05433

Sparrow R. (2016) 'Robots and Respect: Assessing the Case Against Autonomous Weapon Systems', Ethics & International Affairs 30(1), pp. 93-116.

Strubell E., Ganesh A., McCallum A. (2019) 'Energy and Policy Considerations for Deep Learning in NLP', *arxiv.org.1906.02243*. https://arxiv.org/abs/1906.02243

Szegedy Ch. et al. (2014) 'Intriguing properties of neural networks,' *arxiv.org.1312.6199v4*. https://arxiv.org/abs/1312.6199v4

Our Digital Mirror

Erich Prem

Abstract The digital world has a strong tendency to let everything in its realm appear as resources. This includes digital public discourse and its main creators, humans. In the digital realm, humans constitute the economic end and at the same time provide the means to fulfill that end. A good example is the case of online public discourse. It exemplifies a range of challenges from user abuse to amassment of power, difficulties in regulation, and algorithmic decision-making. At its root lies the untamed perception of humans as economic and information resources. In this way, digital technology provides us with a mirror that shows a side of what we are as humans. It also provides a starting point to discuss such questions as who would we like to be – including digitally, which purpose should we pursue, and how can we live the digital good life?

For Antoine de Saint-Exupery (1939), airplanes can become a tool for knowledge and for human self-knowledge. The same is true for digital technologies. We can use the digital as a mirror that reflects an image of what we are as humans. And when we look closely, it can become an opportunity to shape who we are in ways that make us more attractive.

It is perhaps surprising to talk about attraction in the context of digital humanism. Much of its discourse to date is about the downsides of digital technologies: human estrangement, monopolistic power, unprecedented surveillance, security attacks, and other forms of abuses. However, the discourse of digital humanism is not entirely negative. Its proponents believe in progress including by technological means. It is a constructive endeavor and acknowledges that many good things have come from digital technologies. Still, digital humanism calls for better technology and an improved way of shaping our future with the help of digital tools. It demands technologies shaped in accordance with human values and needs instead of allowing technologies to shape humans in often unpredictable and undesirable ways.

E. Prem (✉)
eutema GmbH and Vienna University of Technology, Vienna, Austria
e-mail: prem@eutema.com

However, regarding digital humanism, a program of only technical advancement would fall short of its real ambition. Although technologically driven and proceeding with the intention of improving technology, digital humanism more often problematizes science-technology relationships, human attitudes, and fundamental concepts that we have come to take for granted. In digital humanism, the digital becomes the mirror that impresses a picture of who we are on ourselves and lets us realize what we should change – about us, about technology, and about the lives that we live.

The pervasiveness and ubiquity of information technology means that just about everything that is subject to digitization is also subject to its transformational powers. Central concepts of our lives – both private and professional – have become transformed when they seemingly were only digitized. Collaboration is now something happening in shared directories, meeting new people is mediated by apps and preference settings, news has become pop-up windows, and museums curate data. Intelligence is something that machines exhibit, and chess has been turned into a formal calculation task rather than a board game since around the 1970s. It is as if we had to reinvent the whole world in the digital again and examine its values and implications for our lives. A good example is the case of public discourse, certainly also a feat of the new digital sphere.

1   The Example of Online Discourse

The advent of the internet significantly impacted the way we speak in public. From early newsgroups to today's online social networks, this development deserves its own, more thorough investigation. Today, public discourse without the digital has become quite unthinkable. At the same time, the phenomenon of online discourse is now a major concern of policy makers as well as of critical thinkers, researchers, and many citizens. Its shortcomings, from fake news to echo chambers, from foreign political influence to the pervasion of illegal content, are blamed on its digital form, hence, on technology. Some key challenges include the following:

- Platforms exploit discourse to drive user behavior. They can prioritize emotional content over facts, nudge users into staying online, and have become viable ways to influence user behavior including political decisions.
- Algorithms supervise and police user-contributed online content with the aim to detect illegal matter, spot infringements of intellectual property, remove what may be considered harmful, etc.
- There is a massive shift of power over discourse control from traditional rulers of public discourse, such as media, politicians, and thinkers, to digital platforms.
- Discourse in online platforms has proven enormously difficult to regulate by any single country. The only exceptions are through massive investments in surveillance, censorship, and severe limitations of freedom of expression, for example, in China.

- User-generated discourse provides platforms with large amounts of data to learn, build models, predict behavior, and generate profit in various ways including targeted advertising based on behavioral prediction.

These challenges are by no means unique to online public discourse. We find massive shifts of power toward platforms throughout the world of internet business; platforms have generally proven difficult to regulate – not just regarding public discourse; algorithmic decision-making affecting humans happens through a broad range of applications; harvesting data from all sorts of electronic devices lies at the root of surveillance capitalism; and luring users to stay online through emotional targeting happens across a range of online media today. Online public discourse is really but one example, albeit one that is pervasive throughout societies all over the world.

Beyond the listed concerns, digital online discourse seems to affect members of societies in their sense of belonging. The individualized nature of person-targeted discourse, its character of entertainment, and the self-fulfilling quality of opinionated content sharing and creation have severely undermined shared views, collective narratives, and communal perception. It has been suggested that digital discourse only involves a "simulated" public. Earlier analyses of digital culture focused more on the creation of new communities as well as on referentiality and algorithmicity (Stalder 2016) as ways of creating a shared understanding. Today however, discourse moderation algorithms reinforce an individualized monologue in which references serve to propagate an individual's opinions and lead to the often-diagnosed fragmentation of society. Discourse in the digital world thus runs the risk of endangering the common good not necessarily because of attacking any good specifically but because of undermining the concept of the commons. It limits what is shared among people and thus what contributes to forming a societal collective. This is yet another case of digital technologies not only changing human behavior but changing the very essence of key concepts in often unexpected and unpredictable ways.

A recurring topic in digital humanism is that of primacy of agency or who shapes whom: is technology shaping humans or should technologies be designed in accordance with human needs and values? Unfortunately, matters in digital technologies are never so simple, and there is mutual influence of the two spheres. In digital humanism, this phenomenon has been called *co-evolution*. When co-evolution affects basic concepts, such as discourse, it seems futile to repair these fundamental concept drifts and the challenges they create only by mending technologies. Much beyond co-evolution, there is a deeper, more philosophical question to ask. It concerns a matter of choice and decision: How do we want to be as humans? It concerns ethical choices about *the good life digital* as much as it concerns the design of our technosphere. To stay with the example of online discourse, the digital then poses the question of what type of discourse do we want, or perhaps more ontologically, what should discourse be?

Some legislators and platform owners, for example, suggest using algorithms for improving online discourse. The idea is that artificial intelligence removes illegal content, and many seem to suggest that any content that is potentially harmful should be removed. While the former is usually defined in legal texts and practice, the latter is typically ill-defined and lies at the networks' discretion. The ensuing discussions of democratic parliaments and nongovernment think-tanks then concern freedom of expression as a basic or human right, censorship, regulation, etc. (Cowls et al. 2020). While these are important and difficult discussions, a more essential line of thinking is required, namely, the question of what should the essential qualities of online discourse be? It is another typical characteristic of digital technologies that we can rarely do away with them once they have been rolled out. We therefore need to have productive, forward-looking discussions. This can include a debate about how much "harm" a discourse may have to include to be productive, to stimulate, or to provoke. We need to discuss not only formal qualities of discourse, but what should its purpose be, who should partake, and whom should it serve?

2   Scaffolding Discourse

The reasons for challenges of digital technologies do not exclusively root in the fact that they are digital as opposed to analogue, nor do they lie in their ubiquitous nature and the ease with which digital technologies can manage large numbers. The challenges root in how they affect our basic conceptions of the world. Although the technical characteristics are important, there currently is an unprecedented scale of how the digital facilitates commercial gains of a specific character. We mentioned how online discourse provides a basis of targeted advertising, of data harvesting, and for the construction of predictive behavioral models. This exploitation of online discourse lets discussions appear as a resource in the digital sphere. The digital (platform) perspective thus regards human language from the standpoint of observability and predictability. The resulting digital online sphere consists of (mostly) humans that provide linguistic resources and their online presence and of businesses requiring that humans need to be predicted and targeted in advertising. In this discourse, humans become a resource in the digital realm.

Such a resource-focused perspective is not unique to digital technology. As early as 1954, Heidegger suggested that this specific way of letting everything appear as a resource lies in the very nature of modern technology (Heidegger 1954). In his terminology, technology lets everything become part of an *enframing* ("Ge-stell") as resource ("Bestand"). Heidegger uses the example of the river that appears as a power source once we start building electricity stations. Digitization not only provides such enframing for various objects and phenomena in our environment; it additionally and much more than previous, older technologies enframes us as humans. It is perplexing that in the digital realm, the human is both the source and the sink. Humans constitute the economic end and at the same time provide the means to fulfill that end. Humans stand reserve to the extent that they are simply either data or money generators. From an economic viewpoint, Zuboff (2019) identified a similar concept drift underlying surveillance capitalism. It is a strong and impactful understanding of humans driven by the commercial online world. It is commercially attractive and promising with its double take on the human as a resource.

Engineering may always imply "a certain conception of the human" (Doridot 2008), but we have choices. For example, not all public discourse needs to take such an instrumentalist turn. Like other human activities, we can choose the purpose of our speaking. Some forms of public online discussions are designed to facilitate dialogues among groups of highly engaged speakers of a local community (e.g., MIT's Local Voices Network – https:lvn.org). Others are solution- and goal-oriented and live a practice of focused contributions (e.g., the business and employment-oriented network LinkedIn). Such examples suggest that there are ways to facilitate online discourse less prone to filter bubbles, echo chambers, fake news, etc. and perhaps even in business settings. It also shows how purposes can be designed in line with human needs; in fact, purposes are entirely human made.

We may still have to focus on technology, to occasionally *retreat* from social media, *reform* its way of working, and exert *restraint* as Deibert (2020) suggests. However, realizing that some types of online discourse emerge from the instrumentalization of users, turning them into targets and exploiting them as resources, means to understand not just technology, but ourselves. Technology then is the mirror that presents us with an image of ourselves that we may not find entirely attractive. We can also find possible relief in Heidegger, who quotes Hölderlin: "But where danger is, grows the saving power also." This suggests that the enframing also reveals truth upon which we can build if we want to overcome present danger. And lifted back to the level of digital humanism, the questions then become who would we like to include digitally, which purpose do we pursue, and how can we live the good digital life? Finally, we will also have to find good answers to the question who is "we"?

References

Cowls J., Darius P., Golunova V., Mendis S., Prem E., Santistevan D., Wang W. (2020). Freedom of Expression in the Digital Public Sphere [Policy Brief]. Research Sprint on AI and Content Moderation. Retrieved from https://graphite.page/policy-brief-values

De Saint-Exúpery A. (1939) *Wind, Sand und Sterne (Terre des hommes.)* Düsseldorf: Karl Rauch.

Deibert R.J. (2020) *Reset.* Toronto: House of Anansi Press.

Doridot F. (2008) Towards an 'engineered epistemology'? Interdisciplinary Science Reviews, 33:3, 254-262, DOI: https://doi.org/10.1179/174327908X366941.

Heidegger M. (1954) Die Frage nach der Technik. (The question concerning technology.) *Vorträge und Aufsätze. (1990)*, Pfullingen: Neske.

Stalder F. (2016) *Kultur der Digitalisierung. (Culture of digitization.)* Frankfurt/Main: Suhrkamp.

Zuboff S. (2019) Surveillance capitalism. London: Profile books.

Part IV
Information Technology and the Arts

Fictionalizing the Robot and Artificial Intelligence

Nathalie Weidenfeld

Abstract This text explores the contemporary fascination with robots and digitality and points out how this distorts our view on what digitization can do for us. It pleads for a realist and non-fictionalized view on robots and artificial intelligence.

Nothing in our contemporary popular culture, so it seems, has been more fascinating than the fantasy of robots and the idea of an upcoming total digitization of our society. Hollywood is full of blockbuster films filled with evil, sometimes beautiful robots and dreams of eternal or at least alternative lives made possible through digital means.

Films however are not only expressions of deep-seated fears and hopes, they also create a cultural imaginary that feeds into these fears and hopes, thereby creating a loop that is best described as a more or less closed circle. Now, artistic creations are and should of course be free to do many things; they may create unrealistic settings and invent dramatic premises which make us wonder. Problems only arise when readers or viewers forget to read these films correctly, that is, *metaphorically*. Take a film about a society in which robots are used as personal slaves for household chores, where the protagonist must learn to overcome his prejudice toward them robots. Does this film incite us to think about our relationships to future robots? No! Because this film is not about robots but a metaphorical tale about humans dealing with humans in which robots represent underprivileged humans.

Digitalization and artificial intelligence pose many problems for society and culture. It is therefore of utmost importance to see them for what they are in order to judge their potentials and their dangers realistically. An inadequate "import" of fiction into reality is useless and unproductive.

In order to see more clearly what exactly has been imported from fiction into reality, let's take a closer look at narratives focused on AI. When we look at films – particularly the ones in the last 20 years – dealing with the topos of the robot, we can

N. Weidenfeld (✉)
Munich, Germany
e-mail: weidenfeld@nida-ruemelin.de

97

discern two types: the good, innocent, and sometimes even spiritual robot and the bad, demonical, and evil robot. These two stereotypes are an expression of a paradigm that can be called "primitivist." The primitivistic paradigm is a "cultural reflex" of Western society in need to construct an "Other" that can then be used as a mirror (Torgovnick 1990): a mirror onto which one can project one's own beloved or hated properties. American Indians have long served as Other within the primitivist order – not only in the time of the European Enlightenment but also in the US American culture: During centuries, Indians were either portrayed as bloodthirsty demonic savages or as innocent and spiritually superior people. The American Indian as a topos remains an obsession for US American novels and films, starting with narratives in the seventeenth century dealing with Puritan settlers abducted by bloodthirsty Indians, all through the nineteenth century where narratives of the noble Indian become popular up to the New Age image of the spiritually and morally idealized Indian.

Societies not only create an imaginary Other from already existing real persons but also create them from time to time. The best example for this is the Alien, who became a primitivist topos in the 1980s. The Aliens were portrayed along the same lines as the Indians beforehand: They were either evil and bloodthirsty or good and spiritually highly developed (Weidenfeld 2007a, b). Today, the robots have taken over the role of the Aliens. This primitivist mode of conceptionalizing robots has a deep influence on present society. If Elon Musk or even Stephen Hawking warns us and speaks of lurking dangers and threats which robots pose for humanity – they serve the same old primitivist cliché of the evil Other. What they are doing is not an adequate or realistic description, but rather an *inscription* into an already existing narrative which is at the same time re-introduced into the world.

When one looks into narratives and images of digitization, one encounters similar mechanisms. Digitization is either seen as a mode of bringing about a digital paradise or digital hell. The idea of a digital heaven is often invoked by images of clarity, a blue sky, and an overwhelming universe of all-connectedness. Digital heaven is a puritan heaven: It is based on the idea of transparency, clarity, and purity. In a digital universe, things are either "0" or "1"; there is no ambiguity. Also, the idea of a "natural" teleological development is often suggested in the visual imagery which tells us that man has evolved from ape, to homo sapiens, to homo digitalis – often represented as man holding a portable phone in hand. Digital future becomes a millenaristic prophecy filled with utopian hopes and desires such as the desire for salvation and eternal life.

In the past centuries, technological developments have often been accompanied by unrealistic utopian visions. When Henry Ford pioneered in the mass production of the automobile, he was convinced that this new technology would bring about peace and prosperity for all. "We are about to enter a new era because when we have everything, freedom and paradise on earth will come. Our new way of thinking and acting will bring us a new world, a new heaven and a new earth, as prophets have longed for since age old times" (Ford 1919). Roughly a hundred years after the automobile, it is digitization that will supposedly bring salvation. CEOs in Silicon Valley upload their presentations with images which suggest this.

Digital hell is represented mostly in three possible modes: It is either a digitally economized hell like the world of *Bladerunner* or *Minority Report*, where digital means are used for advertising; or a fascist world, where one ruler has used digital means in order to erect not only his reign but also uniformity and absolute synchronicity; or at least a nightmarish digital world such as in the film *Matrix*, where humans have lost all autonomy and self-determination.

Utopian and dystopian fantasies have moved out of the realm of films and novels into our daily expectations. These religiously motivated expectations take much energy away from that, what is important: discussions on and clarifications of concrete ethical problems that digitization poses.

Lastly, there is a psychological aspect which Rudolf Dux named the "Frankenstein complex" (Dux 1999): the dream of creating a lifelike being that can feel and think like us. This animistic dream, which Freud described so well in *Totem and Taboo* (Freud 1974), is part of a regressive fantasy, similar to a child's wish to fill a doll with life in order to feel powerful and/or not so lonely anymore.

From the student Nathanael in E.T.A. Hofmann's tale "Der Sandmann" up to the visions of femme fatale robots such as AVA in *Ex Machina*, the dream remains the same: a regressive dream of humanlike companionship which is never quite humanlike but humanlike enough to make us dream of whatever we desire: total love, total communion, or total friendship.

"Get a relationship" is the slogan of sex robot critic Prof. Kathleen Richardson who argues against the development of more and more sophisticated lifelike female robots who take humans' attention away from what is really important, namely, relationships with each other (Richardson 2022).

Digital humanism puts *human* relationships first and aims at thinking and using digital technologies instrumentally (Nida-Rümelin and Weidenfeld 2018). Digital tools were neither made by demons nor angels, but *by* humans *for* humans. It is time to keep this in mind and stop introducing fictional imaginary and topoi in our everyday language and discourse. We need a clear and realistic judgement, unclouded by dreams, projections, and regressive fantasies.

References

Dux, R. (eds.) (1999). Der Frankenstein-Komplex. Frankfurt: Suhrkamp.
Ford, H. (1919) Philosophie der Arbeit. Dresden, p. 47.
Freud, S. (1974). Totem und Tabu. Frankfurt: Fischer Verlag.
Nida-Rümelin, J. and Weidenfeld, N. (2018) Digitaler Humanismus: Eine Ethik für das Zeitalter der Künstlichen Intelligenz. Munich: Piper.
Richardson, K. (2022) Sex Robots: The End of Love. Cambridge: Polity Press.
Torgovnick, M. (1990). *Gone Primitive*. University of Chicago Press, 1990.
Weidenfeld, N.. (2007a) Entführt von Außerirdischen: Abduction Narrative als moderne.
Weidenfeld, N. (2007b) Entführt von Außerirdischen: Abduction Narrative als moderne n: Abduction Narrative als moderne Erscheinungsform puritanischer Kultur. Saarbrücken: Südwestdeutscher Verlag.

How to Be a Digital Humanist in International Relations: Cultural Tech Diplomacy Challenges Silicon Valley

Clara Blume and Martin Rauchbauer

Abstract Digital humanism is often seen as an antidote to the excesses of Silicon Valley and its underlying cultural values. It is however very short-sighted to label big tech exclusively as a threat to our humanistic values, since it has proven to be an essential ally, particularly in the context of the ongoing digital transformation of the international system and its negative impact on human rights and privacy. The emerging field of cultural tech diplomacy has established a new meeting point in the center of global innovation between diplomats, policy makers, artists, and technologists in order to positively shape the future of technology according to our needs and our full potential as human beings. A new digital humanism empowered by artists can serve as a compass for diplomats and technologists alike to serve their citizens and customers while navigating a world radically transformed by artificial intelligence and biotechnology.

Technology[1] has become the most pressing question of our time. Who creates it, who controls it, who has access to it, and who doesn't are the new parameters that determine emerging power structures around the world. Until recently, most of these decisions were taken unilaterally by a handful of tech entrepreneurs in Silicon Valley, and supported through the laws of a free and unregulated market, the obscure inner workings of which regulators around the world neither understood nor wanted to inhibit. Built upon this nurturing humus, Silicon Valley has experienced a long period of economic boom fueled by advances in technology. Globally admired for its ability to attract a seemingly limitless influx of human talent and venture capital, the region has become a symbol for youthful utopian optimism and

---

[1] "Technology" in this article refers to frontier technologies such as artificial intelligence, quantum computing, blockchain, biotechnology, and a.o. that are considered main drivers of the Fourth Industrial Revolution and have created a diverse ecosystem of technology companies. Information technology (IT) companies are a subset of these tech companies.

C. Blume · M. Rauchbauer (✉)
Open Austria, San Francisco, CA, USA
e-mail: clara@open-austria.com; martin@open-austria.com

© The Author(s) 2022
H. Werthner et al. (eds.), *Perspectives on Digital Humanism*,
https://doi.org/10.1007/978-3-030-86144-5_15

techno-utilitarianism. It has also created a powerful culture that believes in the necessity to disrupt existing and to create new markets, in up-scaling for exponential growth, as well as in the idea that technology can solve any problem in the world without the need of government intervention (O'Mara 2020, pp. 30–32). In many ways, Silicon Valley has become a pop-cultural phenomenon, galvanizing the dreams and aspirations of an entire generation.

However, the euphoria and exuberance of Silicon Valley's tech boom has recently shifted. Across the political and ideological spectrum, big tech companies such as Google, Facebook, Apple, Microsoft, and Amazon are increasingly viewed with suspicion for their unprecedented influence and size that in many markets amounted to monopoly power. The sum of many troubling incidents constituted the swelling tide that led to a global decline in Silicon Valley's popularity as well as a growing call for governmental intervention and the return of the nation-state. 2018 was a watershed moment for the tech industry, which eroded the industry's self-confidence. It was in that year that Facebook's CEO Mark Zuckerberg first had to testify in front of US Congress and explain to seemingly ill-prepared lawmakers why the company was selling its users personal data and followed a business model that relied on targeted advertising. In the European Union, the admiration for Silicon Valley around the same time gave way to a desire to reign in big tech, particularly in conjunction with its impact on basic rights and privacy. In May of 2018, the EU's General Data Protection Regulation (GDPR) entered into force, a comprehensive legislation that provided European citizens and tech companies with clear guidelines for the protection of their online privacy rights. At around the same time, the world was realizing how social media platforms could have a role in aiding genocide, when the military in Myanmar used Facebook to launch a ruthless misinformation campaign that targeted the country's Rohingya Muslim minority group. Technology had lost its innocence. And governments, particularly in Europe, felt emboldened. The end of the era of self-regulation gave way to a new form of dialogue between tech companies and nation-states, based on a new humility and the possibility of mutual respect and understanding.

But the current crisis around technology has also laid bare how unequipped most regulators, politicians, policy makers, and diplomats are to engage in a conversation with the tech industry on equal footing. In international affairs, even the most digitally literate diplomats and government officials find themselves often ill-suited to navigate all aspects of digitalization and its effects on citizens. New technologies have impacted our world in a much more profound and permanent way than foreign policy institutions often seem to be aware of. Traditional fields of diplomacy such as geopolitics, security policy, international law, multilateralism, human rights, development cooperation, consular affairs, as well as cultural diplomacy could benefit from some of the innovative spirit that laid the groundwork for Silicon Valley's success story in order to keep up with the increasing pace of emerging technologies. If they don't adapt, new forms of cyber warfare and advances in artificial intelligence will increasingly create unprecedented vulnerabilities for entire countries and the international system as a whole.

Communication is key. But it often seems as if technologists and policy makers don't share the same language. Language is coined by cultural context, professional backgrounds, education, mindset, and the complex assembly of layers that constitute each human being shaping our understanding of the world around us. So when a Silicon Valley technologist uses terminology such as "technology," "truth," "freedom," or "humanity," they don't necessarily attribute the same meaning to that same word uttered by a European policy maker. But these nuanced distinctions and conveyed subtexts are crucial and need clarification in order for policy makers to work on the human project in the years to come. We're in dire need to develop a new vocabulary that works as a universal language equipped to address the question of what it means to be human in the digital age. But the main reason for the linguistic and cognitive dissonance between Silicon Valley and the rest of the world is often rooted in its contrasting vision for the future of humankind.

Silicon Valley's tech pioneers today are often unaware of the long-standing roots of their famed mindset. Some characteristics go way back to the frenzy of the historic gold rush, aligned with the American dream of global exceptionalism and the urge to push the Western frontier. The dogma of absolute freedom of information and the aversion against any central authority preached by many internet pioneers can be traced back to the San Francisco-born hippie counterculture of the 1960s. But the tech industry has been shaped similarly by a strong belief in self-reliant entrepreneurship (Markoff 2005). Together with stunning technological and scientific breakthroughs, these diverse attitudes merged into the unique Silicon Valley ideology of transhumanism, inspired by tech evangelists like Raymond Kurzweil (2005) and Hans Peter Moravec (1998). For a transhumanist, death and old age are seen as mere limitations that can be cured through biotechnology and robotics. There are no limits to the human project, when genetic engineering paves the way for self-optimization of one's "flawed" DNA. Even our dying planet can be seen as merely another station along humanity's voyage through the universe. For some of Silicon Valley's elites, to "go west" can mean, quite literally, to reach for the stars. The leading tech pioneers like Elon Musk, Sam Altman, and Jeff Bezos are therefore spearheading a new elite that wants to break this final frontier, but only for a selected few with the financial means to embark on the journey. The mythology of the transhumanists is influencing how technology is being shaped for the rest of humankind. But what is needed now more than ever is a common vision for a digital humanity that is not exclusively determined by a handful of individuals with a ticket to Mars. And while this is attached to a new host of problems, humanity's best way forward is to forge an alliance with tech companies that align around a certain set of humanistic values.

The concept of digital humanism can be seen as a necessary evolution of Silicon Valley's transhumanist vision for humanity: a marriage of transhumanists' excitement about the immense potential of technology and a new humanism that aims to restore our dignity online and offline. A digital humanist understands technology as a tool that can be used for both good and bad, realistic about its potential to elevate us and cause harm though a host of unintended consequences (Nida-Rümelin and Weidenfeld 2018). Equipped with a certain pragmatism that is tilting neither toward delusional techno-utopianism nor fatalistic fear-mongering, tech policy making and international relations based on a new digital humanism want to create and advocate for global frameworks for technology that preserve our universal human rights.

International diplomacy is practiced through its own language code. In multilateral fora such as the United Nations, where new ethical principles, norms, standards, and even instruments of international law are negotiated, different nations rely on different words to camouflage their underlying agenda and sometimes to embellish reality. Authoritarian states, for instance, rely heavily on the words sovereignty (Applebaum 2020) and non-interference to push back when pressed on recent online human rights violations. It's through these dog whistles that human rights are increasingly under attack, not only in practice but also conceptually. We are also facing an increasingly polarized and fragmented world in danger of creating parallel ideological universes, conceptual frameworks, and a separate digital infrastructure for new technologies (Lee 2018). Against this emerging geopolitical divide, diplomats often struggle to address common concerns that are based on our shared humanity. Digital humanism lends itself to become a universally accepted concept for international diplomacy that can rally countries with different political systems, cultures, and histories around a certain set of values without compromising on universally recognized human rights. As a tech policy compass, it could guide nation-states and tech companies to work together on the digital transformation of our international system. Digital humanism could be the safe space where even ideological adversaries and competitors could find common ground. It is in everyone's interest to develop technologies that are trusted by consumers around the world. Even digital authoritarian countries, which engage in an arms race for global influence and power, have no interest in creating rogue autonomous weapons systems that they themselves may one day lose control over. Tech companies are also in dire need for a compass that values culture over strategy as a necessary environment for innovation and growth. Digital humanism could very well provide the tool kit for a new corporate culture of big and small tech companies and provide orientation in times of global tectonic shifts.

Digital humanism can also serve as a blueprint for nation-states that want to engage with the global tech industry. Since some big players in technology boast annual revenues that easily match the GDP of smaller nation-states, diplomacy needs to rethink what it means to be an international actor in the digital age. One such innovative approach in international relations is the emerging field of tech diplomacy, applied by an increasing number of nation-states and other actors in Silicon Valley. The Danish initiative of "techplomacy" was a milestone for international diplomacy and received global attention. In 2020, Austria followed suit as the second country in the world to appoint a Tech Ambassador to Silicon Valley. To mediate between governments and big tech, tech diplomacy addresses an imbalance in information and competence. Traditionally, tech companies have lobbied for their interests in political centers of influence. Today, governments are sending diplomats to lobby for the interests of their citizens to the technological centers of influence. Of course, tech diplomacy is reciprocal and therefore also attractive for tech companies. They can see their prestige and influence on the international system formerly recognized by being included into a practice traditionally only reserved to sovereign governments. But tech companies have other incentives as well. They employ their own tech diplomats to show to the world that their long-term interests go beyond the immediate goal of making a profit. One might argue that their business depends on a stable rules-based international system and is strengthened by a shared set of values and their technology being developed through the lens of common ethical principles.

Tech diplomacy is only one way that governments and tech companies interact with each other. CEOs of big platforms often meet and consult with government leaders directly. But the impact of these encounters seldomly trickle down to the institutional level of government, lack continuity, and often a clear agenda. Tech diplomacy, on the other hand, institutionalizes the relationship between the tech industry and governments by creating long-lasting relations and mechanisms that can be activated in times of crisis and need. To be clear, tech companies are not states. However, in some areas, their growing power challenges traditional realms of government (Wichwoski 2020). Sovereign countries may control their territories, but big tech platforms control the "digital territories" of their online communities and can define the rules which have a spillover effect on public life in general. Tech diplomats need to function as advocates of their citizens and make tech companies accountable. However, tech diplomacy is also about seeking alliances and common ground where interests and values align. In this context, a new digital humanism in diplomacy can also lead democratic governments and tech companies who share a set of principles and beliefs to collaborate in the fight of digital authoritarianism globally.

In addition, foreign policy actors need to take into account that there is an inherent difference between human beings and their digital avatars. The consequences of merging the real and online world, while we gradually turn into digital humans, aren't clear yet. In order to understand the many layers that constitute a digital human policy makers, tech companies and technologists need to break out of their respective silos and start working together to assure that basic rights are met regardless of these differences. The novel interdisciplinary concept of cultural tech diplomacy was first pioneered by Open Austria, the official Austrian government representation in Silicon Valley. Traditional cultural diplomacy uses art and culture as a soft power tool (Nye 1990, pp. 153–171) to influence other nation-states and their citizens by means of attraction rather than coercion. Think about the successes of Hollywood, Rock 'n' Roll, and David Hasselhoff inspiring young Germans to "look for freedom" that helped to end the Cold War. By combining the immense potential of cultural soft power with tech diplomacy, the Austrian diplomatic mission in Silicon Valley creates a safe space for dialogue with tech partners that is rooted in artistic experimentation. Art lends itself to explore topics that in other arenas would be prone to conflict and trigger political controversies and partisan agendas. To be the "fool" who is able to speak the truth has a lot of merit, especially in the delicate inner workings of the international system.

A recent study commissioned by art + tech network The Grid led by Open Austria, highlights the asymmetry of power between artists and tech companies who are limiting the access to their technologies. But this asymmetry "obscures the foundational role that the region's counterculture has played in the emergence of big tech – and the values of techno-utopianism, flattened hierarchies, flexibility, and so on that have guided the industry," as the report's author Vanessa Chang (2020) points out. Artists have historically contributed greatly to Silicon Valley's world-conquering success formula. It's not only fair, it's imperative that artists should yet again be included in the creative process that constitutes the development of new tech. It is essential to reassign value to artistic practices inside the technology sphere. Complementary to the top-down regulatory approach that reacts to existing technology, cultural tech diplomacy wants to proactively shape technology bottom-up during the conception, research, and development of new tech products and services. For this model to be successfully implemented, tech companies need to make themselves vulnerable to the arts by opening up their R&D labs to artists and philosophers that can provide a whole new perspective to an old set of problems. Artists as digital humanists are uniquely equipped to explore the potential and pitfalls of frontier technologies in an unconventional and experimental way that has the additional benefit of not only educating technologists but also the broader population about what it means to be a digital human in the age of artificial intelligence.

At a time of shifting core values, an acceleration of digitalization, and a changing international system, diplomats today need to be agile and transform into digital humanists in order to successfully navigate the challenges of their trade. Cultural tech diplomacy could be a winning proposition that paves the way for the future of international relations in the digital age.

References

Applebaum, Anne (2020), *How China outsmarted the Trump administration. While the U.S. is distracted, China is rewriting the rules of the global order*, Washington DC: The Atlantic, Nov 2020

Chang, Vanessa (2020) *The Grid: Art + Tech Report 2020*, San Francisco : EUNIC Silicon Valley, December 2020

Kurzweil, Ray (2005), The Singularity Is Near: When Humans Transcend Biology, New York: Penguin

Lee, Kai-Fu (2018), AI Superpowers: China, Silicon Valley, and the new world order, Boston: Houghton Mifflin Harcourt

Markoff, John (2005), *What the Dormouse Said: How the Sixties Counterculture Shaped the Personal Computer Industry,*' New York: Viking Press

Moravec, Hans (1998), Robot: Mere Machine to Transcendent Mind, Oxford University Press.

Nida-Rümelin, Julian, Weidenfeld, Nathalie (2018), Digitaler Humanismus - Eine Ethik für das Zeitalter der Künstlichen Intelligenz, München: Piper Verlag GmbH.

Nye, Joseph S. Jr (1990) *Soft Power*, Foreign Policy, Nr. 80, Twentieth Anniversary, Autumn 1990

O'Mara, Margaret, (2020), *Silicon Politics,* Communications of the ACM, December 2020, Vol. 63 No. 12

Wichwoski, Alexis (2020) The information trade: how big tech conquers countries, challenges our rights, and transforms our world, New York: HarperCollins

We Are Needed More Than Ever: Cultural Heritage, Libraries, and Archives

Anita Eichinger and Katharina Prager

Abstract Libraries and archives as institutions of cultural heritage have a long history of and great expertise in collecting, securing, handling, and contextualizing masses of material and data. In the context of digital humanism, these institutions might become essential as a model as well as a field of experimentation. Questioning their own role as gatekeepers and curators, the digital transformation offers them the chance to open up – through both participatory initiatives and inclusive collecting. At the same time, however, it is a matter of preserving the library and archive as a place of encounter and personal dialogue in a human and humanist tradition.

The Corona crisis has given digitization a huge boost in many areas, which need to be developed further in a strategic and co-designing way – without subordinating ourselves to technologies but shaping them. The City of Vienna currently encourages and orchestrates innovative ideas, collaborative action, and social engagement in this field in its framework initiative "digital humanism." With Anita Eichinger, the Vienna City Library was involved from the very beginning and took on the mentorship of the working group on arts and culture.[1]

The Vienna City Library,[2] often described as "memory of the city," has developed over the last 170 years from an inconspicuous administrative library (founded in 1856) into a representative city library, a municipal cultural institution, and one of the most important scholarly archives with a special focus on Vienna. It holds around 6,500,000 volumes of books on Vienna (Viennensia and Austriaca); 1600 bequests of important inhabitants of the city (such as Franz Grillparzer, Marie von Ebner-Eschenbach, Franz Schubert, Johann Strauss, Karl Kraus, or Friederike Mayröcker),

---

[1] It consists of Gerald Bast, Daniel Löcker, and Carmen Aberer (MA 7) and Irina Nalis, Elfriede Penz, Erich Prem, Eveline Wandl-Vogt, and Hannes Werthner as well as Anita Eichinger and Katharina Prager (MA 9) and formulated a position paper on "Digital Humanism and Arts/Culture."

[2] See https://www.wienbibliothek.at/english and https://www.digital.wienbibliothek.at/.

A. Eichinger · K. Prager (✉)
Vienna City Library, Vienna, Austria
e-mail: anita.eichinger@wienbibliothek.at; katharina.prager@wienbibliothek.at

© The Author(s) 2022
H. Werthner et al. (eds.), *Perspectives on Digital Humanism*,
https://doi.org/10.1007/978-3-030-86144-5_16

some of them part of the UNESCO Memory of the World; and one of the world's largest poster collection. Apart from these central holdings, it also collects newspaper clippings, historical photographs, sermons, leaflets, travel reports, cookbooks, and much more. Over the past decade, there has been a strong focus on retro-digitization – and while digital accessibility of materials remains a priority, the library is concentrating on research, innovation, and digital humanism in the years to come.

Although the Vienna Library is not an art and cultural institution per se, it is an important interface in the cultural field, with expertise in safeguarding and mediating cultural assets.

In our current global situation – where we are dealing with the monopolization of the web, the spread of extremist attitudes, anti-factualism, filter bubbles and echo chambers, the loss of privacy, and many other problems – the importance of archives and libraries cannot be overstated. Historian Jill Lepore most recently pointed out these connections and the history of evidence, proof, and knowledge in her podcast "The Last Archive."[3]

Keeping track, filing, and cataloguing are important tools in bibliographic control. However, it is also essential to understand creativity, imagination, social and critical competence, change of perspective, inclusion, diversity, and much more as central contents of cultural and artistic activity. The abovementioned working group on arts and culture made it clear in its position paper that art, culture, and the competences of the creative must help shape digital humanism as fundamental factors – a matter that has been obvious to technology monopolies, for example, for a long time already.

It can be argued that libraries and archives also have some experience in combining creativity with order or chaos with systematics and adapting their practices to the logics of human art and knowledge production over centuries.

The Vienna City Library keeps the historical records of a city renowned worldwide as a city of culture and – in the last decades – also as a center of social, scientific, and technological innovation. In this respect, it can also refer to its legendary intellectual history: Before and after World War I, *fin-de-siècle Vienna* and *Red Vienna* achieved international significance in the cultural and social sphere. After the end of the Austrian monarchy in 1918, the city's leaders, together with its intellectuals, boldly "imagined a new society that would be economically just, scientifically rigorous, and radically democratic. 'Red Vienna' undertook experiments in public housing, welfare, and education while maintaining a world-class presence in science, music, literature, theater, and other fields of cultural production" (McFarland et al. 2020). The roots of the ideas that came to life in the first Austrian republic were often already established in *fin-de-siècle Vienna*. Mostly, they were a reaction to profound societal, technological, and medial changes. The fields of medicine, economics, art, and philosophy reacted to this turmoil and sought for new ways of living – Freud's psychoanalysis, the philosophers of the Vienna Circle,

---

[3] See https://www.thelastarchive.com/.

or the epochal innovations in music driven by Schönberg and the Vienna School are still known to this day. Today, we are in a similar situation of upheaval. Digital humanism aims to encourage people to think and to shape the digital future in a new way. So much for the broader context – but what do digital transformation and digital humanism involve in the context of the Vienna City Library, whose duty it is to preserve the cultural heritage of the city – and also for libraries and archives in general?

Digital humanism calls for a "third way" of digitization. This means that there must be an alternative way to Silicon Valley and China, a way without aiming for profit or authoritarianism, but for the benefit of humanity. In 2004, the Google Books project started. Google worked together with libraries and publishers around the world to preserve books and to make the information accessible to people world-wide. Prominent and huge university libraries have cooperated with Google since that time. Jean-Noel Jeanneney, head of Bibliothèque nationale de France from 2002 to 2007, cautioned against Google and "Americanization" and argued for a European digital library (Jeanneney 2006). Although Jeanneney was polemic in his book, there is one important conclusion to be drawn: Cultural heritage is a public good, therefore, it should remain public property. Digitizing books and other sources in libraries and making them accessible to the public and the scientific world must be the responsibility of public non-commercial institutions. Parallel to Google archives, libraries and museums around the world also started massive programs not only to digitize their collections but also to contextualize them and, therefore, gain additional values and new insights (e.g., citizen science projects, digital history platforms, digital humanities projects). Libraries are at a turning point. They have good prerequisites and qualifications, but they need to change their perspective on what they have been doing for thousands of years. The Vienna City Library takes digital humanism as a chance to reposition itself. In the following, we outline four important new tasks as part of its strategy to answer the challenges of the digital age.

1 Self/Education

This term is meant to combine two aspects, namely, the mission to educate a wider public as well as oneself as part of a library. First, it should be a central task of libraries to systematically counteract to the digital gap and to train critical, responsible users, as well as designers of our digital life, together. Archives and libraries have habitually and for centuries dealt with masses of data, disorder, gaps, and search and selection processes. The systematic, precise handling of data is one of their core competences, and they should learn to impart this knowledge, which is so often a desideratum elsewhere, to a wider audience. In the spirit of digital humanism, this should also increasingly involve marginalized and disadvantaged groups to whom the culturally transmitted knowledge of a community has often not been easily accessible for various reasons (language, educational background, etc.). The role of the librarians and archivists is changing from reclusive gatekeepers of hidden treasures to guides that help users navigate contradictory data systems. In this context, it is necessary for librarians to acquire skills of the future and to try out new cultural techniques. These include, among others, dealing with ambiguity and uncertainty; imagination and association; intuition; thinking in terms of alternatives; establishing unconventional contexts; challenging the status quo; and changing perspectives.

2 Participatory Turn

To get this self/education started and to initiate a "participatory turn" in archives and libraries, the Vienna City Library aims at launching a pilot project in the realm of digital humanism. Under the working title "WE make history," the first step will be to enter cooperations with peer institutions documenting the city's history and to link and visualize all digitized and digitizable sources. In a second essential step, it will then be a matter of inquiring who remains invisible and why. As a result, it is intended to make it possible to enrich, supplement, and retell the city's multi-perspective history, in order to offer groups that have been excluded from the creation of cultural heritage opportunities to contribute their stories and their versions of history. For instance, in the historical address books from 1859 to 1942, only the heads of the households were listed, and they were mostly male. Combining layers of resources and research, we will not only make all the other persons – especially the female half – visible but also take a step further when we ask the Viennese public to enrich these layers of date with life stories, photographs, documents, or other sources. This model project will not only lead to a rethinking of how cultural heritage is conveyed – it will also help to rethink another important area – namely, the question of what a library collects and how.

3 Inclusive Collections

Collections in archives, museums, and libraries – this is frequently stated – are hardly neutral nor objective storehouses, but reflect power relations. The Vienna Library was no exception when it collected the written material published in Vienna over the last 170 years, as well as the papers of famous authors, musicians, and cultural figures – 76% of them of male gender. Digital collecting often appears to be the solution for challenging tradition, opening up, and expanding centuries-old structures, but experimenting with digital collection methods in the context of 9/11, Hurricane Katrina, and most recently COVID-19 has shown that it poses new difficulties and opens up other gaps (Rivard 2014). Digital humanism will be a central starting point for critically reflecting on the interplay between technology and identity politics and collecting material in a correspondingly different way. Inclusive collections are, of course, intrinsically linked to a new self-image of librarians and archivists – from "guardians of the past" to actors who are concerned with the present, the future, and the construction of social memory (see *Self/Education*). Secondly, radical repositioning needs the support and participation of a broad community (see *Participatory Turn*). Altogether, it might be useful to remember that the Vienna City Library always took curatorial liberties when establishing its special collections. It is important to be aware that changes take time and can only be achieved one step after another, project by project.

4 Remaining a Place for Personal Encounters

While navigating through these uncertain times of digital transformation in the spirit of digital humanism by self/education, fostering participation, and reframing our collections, it is helpful to keep our feet grounded and remain physically in place. The closure of the archives and libraries, due to the COVID-19 restrictions, did not only lead to debates about the value of archival research but also caused longing for the special atmosphere of these places. In this context, the reading room is not just an arbitrary workplace – although these, too, were nostalgically transfigured in the lockdown – but a special place of exchange that does not only establish the connection between the material and the researcher but also between the archive and the researcher, between the researchers themselves, and, ultimately, between the current collective issues and the collective memory.

The fundamental question is what will remain of archives and libraries, once it has actually been possible to make cultural heritage digitally and barrier-free accessible to all. Will they dissolve as locations or can they gain new significance as venues for analogue debate and human encounters – and if so, how? This field of tension is opening up by digitization, but digital humanism at least hints at answers as to why spaces of human encounters will remain essential – and the experience of a pandemic confirms this.

5 Conclusions

Libraries, archives, and our overall attitude toward our cultural heritage are at a crucial turning point in times of digital transformation.

On the one hand, the more digital our lives become, the more we need places like libraries to discuss, interact, and invent new innovative solutions together. On the other hand, the profession of the librarian is often perceived as declining and compared to a soon extinct species. The contrary is the case. Without librarians, we would give our cultural heritage and, therefore, our cultural identity out of our hands. The question of what to collect in the future and how to preserve and protect the digital heritage can only be discussed in participatory exchange and cooperation of librarians with (citizen) scientists and – last but not least – computer scientists. In this context, librarians need IT experts to understand which spaces they are opening and closing in the digital realm, how they can position themselves meaningfully at the interfaces, and where data is secure. But IT also needs librarians more than ever to recognize that key practices for dealing with cultural heritage are already in place and that in many cases they can be transformed into the digital realm. The most crucial thing is to realize that all the challenges ahead can only be met through multidisciplinary exchange and mutual understanding.

References

Jeanneney, Jean-Noel (2006) Googles Herausforderung. Für eine europäische Bibliothek. Berlin: Wagenbach
McFarland, R. Spitaler, G. and Zechner, I. (ed.) The Red Vienna Sourcebook. London: Camden House 2020.
Rivard, C. (2014) Archiving Disaster and National Identity in the Digital Realm: The September 11 Digital Archive and the Hurricane Digital Memory Bank, in: Rak, J. and Poletti, A. (ed.) Identity Technologies: Producing Online Selves. Wisconsin: University of Wisconsin Press, pp. 132–143.

Humanism and the Great Opportunity of Intelligent User Interfaces for Cultural Heritage

Oliviero Stock

Abstract In the spirit of the modern meaning of the word humanism, if technology aims at the flourishing of humans, it is of the greatest value to empower each human being with the capability of appreciating culture, in an inclusive, individual-adaptive manner. In particular, in this brief chapter, the case is made for the opportunity that intelligent user interfaces can offer specifically in the area of culture, beyond the obvious infrastructural advantages we are all familiar with. Insight is provided on research aimed at the continuous personal enriching of individuals at cultural sites, approaching the ancient humanistic vision of connecting us to our cultural past, now made possible for all, not just for an elite.

Humanism puts humans at the center of interest for all aspects of life, on a philosophical as well as on practical terms. Its roots are in Cicero's term *humanitas*, which in substance meant the development of all forms of human virtue and became an important movement in Italy in the fourteenth century, including outstanding figures of culture and art, such as the poet Francesco Petrarca, before spreading to other areas in Europe. Humanism emphasized the connection to classical culture and, in a way, offered to overcome limits of time. It was not only passive tribute to ancient culture, but active connection: authors like Petrarca gave meaning to the concept of cultural heritage and went all the way to even write letters directly to classical authors.

I really believe we are now at a historical point, one that can steer the human relation to cultural heritage and other cultural aspects in the spirit of a modern, digital humanism. If technology aims at the flourishing of humans, it is of the greatest value to empower each human being with the capability of appreciating culture, in an inclusive, individual-adaptive manner. In particular, in this brief chapter, I would like to make the case for the opportunity that intelligent user interfaces can offer

O. Stock (✉)
Fondazione Bruno Kessler, Trento, Italy
e-mail: stock@fbk.eu

© The Author(s) 2022
H. Werthner et al. (eds.), *Perspectives on Digital Humanism*,
https://doi.org/10.1007/978-3-030-86144-5_17

specifically in the area of culture, beyond the obvious infrastructural advantages we are all familiar with.

In general, we can say that IT and the Web, though they have offered enormous opportunities for human cultural enrichment, have not met the expectations that many had. Most of us believed that technology would have brought same rights for all, opportunities to advance the cultural level, natural exposition to different points of view, in sum a cultural improvement in our society. It has happened to a very limited extent.

At the same time, undesired effects have been widespread; the digital world has brought with it a strong danger of pseudo-cultures that hide aggressive intents and, on the other hand, of cultural uniformity, not to say cultural imperialism. Often, eventually technology has been used for spreading deplorable contents and has even been the tool of choice for hate messages, without us doing much to prevent it. Only recently, a defensive technological effort has started to counter hate speech, and, just to mention a popular theme, fake news are still not easy to detect and counter automatically nor semi-automatically. So there are many challenges ahead on the digital *defense* side, on the intersection of culture and ethics.

In any way, as I said, here I would like to discuss the great *positive* potential of intelligent technologies for cultural heritage. Cultural heritage has many forms; some forms are meant to be immaterial and reproducible from the beginning, like for instance texts, music, or films and, basically, theater. Intelligent technologies may help accessing and interpreting the material. A good example is natural language processing techniques for automatically finding influence relations in concept formation among different authors (see Van Camp and Van den Bosch 2012) or in determining (causal) chains in historical events. Yet, here I would like to focus on physical cultural heritage, on the incomparable experience of being in front of the original material artifact, being it at museums, at historical or art-relevant sites, or possibly also at other unstructured "every day" locations. For a museum, there are three main aspects for digital intervention: preservation, organization, and appreciation. Artifact preservation by means of technologies has a long history and is improving steadily, but it is not the focus of this chapter. Organization is the classical work of the curator, who may involve architects for getting the best out of the combination of exhibits and available space. In modern museums, technology may help for all aspects of design and basic offer to the public, for security in the museum, and also for real-time sensing of visitors' behavior in order to improve the availability of resources for the visitors.

The really novel prospect – and the one addressing the core of humanism – comes from intelligent interfaces for cultural heritage appreciation by the visitor. In abstract, we can think of three phases: before the cultural visit, during the visit, and after the visit. The "before the visit" phase, obtaining information and getting prepared for making the most out of the coming visit, has already a variety of tools available. In addition, in the future, we shall enjoy continuity, so that the actual visit exploits what was explored earlier at home, including a model of the visitor (see Ardissono et al. 2012), acquired beforehand.

In fact, when it comes to the actual visit, the key element is that nothing should take the place of the emotional experience of being in front of the original artifact. The computer interface to culture, and in general any interface, must not pose limits to a natural experience but should augment it. In addition, the computer-based intervention should connect the current experience to a learning model. We want the interface to take into account the cognitive, the emotional, and the physical state of the visitor (where he/she is, tiredness, etc.), to be guided by his/her own motivation, tastes, and preferences, in case, to negotiate what should not be missed, but not to impose a rigid agenda. This flexibility and user adaptation require the visitor model, which must be as accurate as possible, to be continuously updated in the course of the visit. Presentation strategies then depend on media and modalities available, but again should take into account movement in space, and what was explored previously, so that the presentation, which necessarily is language-based, is appropriately personalized and can refer to what drew the visitor's attention previously (Kuflik et al. 2011). Various techniques have been studied, based on hearing, combination with reading, combination with seeing images, all the way to dynamic documentaries with a multimedia narration produced automatically for the current situation (Stock et al. 2007). It is also only natural in this context to be able to provide cultural information from different points of view. Cultural descriptions may be controversial, and criticism and diverse viewpoints add to our understanding. On the visual side, various forms of augmentation of what is being perceived have been proposed, for instance, reconstruction of a building superimposed on the view of the existing fragment (Aliprantis and Caridakis 2019).

People tend to visit museums and historical sites in small groups, family, or friends. Etnographers have shown that conversation is a fundamental factor in the success of the cultural experience. Sharing the emotion, discussing, criticizing, enlarging the perspective help going deeper, learning, and developing a taste for the cultural experience. Also for this aspect, intelligent interfaces can help. Just to mention one example: while theater in the museum had been proposed for some time, inspired by a mobile theater tradition that we can date to the Middle Ages, an original smartphone-based drama technology for the museum was recently created and initially experimented with. The intelligent technology-based drama system gives an active role to visitors and subtly induces conversation about the exhibit contents among the group members, while they move along in their visit (Callaway et al. 2012). This approach is based on dynamically adapted scenes and requires as enabling technology, in addition to accurate positioning, also proximity and group behavior detection. It involves distance communication, and it can be exploited for allowing participation to the visit by elderly or handicapped members of the small group, who cannot leave home.

Another aspect that technology consents is some form of interaction across time: for instance, leaving traces of a visit, in the form of spoken comments that could be heard by your grandchildren when they will happen to be just at the site you are visiting now. Or, more sophisticated: entertaining a dialog with someone not there anymore, through interpretation of visitor's utterances by a dialog management system and clever understanding and composing interview fragments of the departed one (Artstein et al. 2014).

The "after the visit" setting is important for group sharing and reflecting about the visit and consolidating the group experience. At this point, it is obvious that intelligent technology, having a record of the users' competence and current experience, of what each one has seen, what drew his/her attention, etc., can help the conversation, help reinforcing the memory, and provide new stimuli for the individual and for specific insights.

So much for the visit to the museum, or to the art site, as the focus point of "material" culture. A challenge for the future is also to connect *all* cultural experiences. The idea is in the first place that a system that accompanies you to a visit knows about your previous visits to the present museum or to other ones, about what attracted you; it may have a model of how much you may recall and what the novel knowledge should be integrated to. More ambitious is the idea of ubiquitous cultural experience: in all circumstances, the fact that you are nearing a certain site may lead a proactive system to negotiate with your individual model, so to promote some local cultural aspects related to your interests, and find the best way to get your attention and the time for exploring the site and having a personalized presentation (Kuflik et al. 2015). In this spirit, for instance, sites of historical events can be connected to what was learned in a museum, or history narration can be expressed not only in relation to locations of big events but also for "bottom-up" history. To complete this picture, it could be up to local residents, and especially for school projects, to design contents to value their territory.

Having spoken of the opportunity for cultural heritage appreciation, I would like to mention a different, but socially important, theme, still related to cultures, in this case mostly meaning ethnical aspects. I refer to the proposition of technology to facilitate overcoming a conflict. Attention has been given to technology for helping solve conflicts by addressing the basic needs of the two sides, in this way supporting decision-makers. Yet, there is a fundamental question concerned with laypeople involved in a conflict, a question of recognizing the other and shifting attitudes. Intelligent experimental systems have been designed to facilitate the joint creation of a narrative acceptable by participants to the conflict, and studies have been conducted showing that the experience with such systems can help change the attitude toward the other (Zancanaro et al. 2012).

A final note is about ethics in interfaces. In most situations I have tried to describe, the key goal is to motivate individuals and have them find pleasure and interest in going deeper into cultural heritage. Even more obvious is the case of group activity, including the last described goal of nudging participants to the shared narrative. The question of which subtle means for influencing and for nudging through the interface are ethically acceptable must be posed for interfaces and communication. Ethical studies on acceptability of machine persuasive communication (Stock et al. 2016), possibly taking into account different cultures, are an important research theme to be pursued.

In conclusion, I think that we have an extraordinary opportunity with the affirmation of intelligent user interfaces for cultural heritage appreciation. They require fundamental interdisciplinarity: they are based on AI and engineering, but give a central role to cognition, to emotion, and to studies about aesthetics and social sciences. In addition, of course, to all the wealth of contents concerned with culture itself.

At the time of the original fourteenth-century humanism, of course the view of culture was just for a small elite. Many centuries later, intelligent digital technology can offer previously unthinkable means for greater appreciation of culture and also extraordinary flexibility: it can help everyone, from the cultural expert to the newcomer, call out his/her inner, human attitude toward learning, toward beauty, and in general toward knowing our (and others') past, potentially thus facilitating the understanding of all that is human. This opportunity is not for the elite; it is for all, in the spirit of the modern meaning of the word humanism. Original humanism was the initial step that led to what came to be known as Renaissance. Will the opportunity of digital humanism be well understood and lead us to a digital Renaissance?

References

Aliprantis, J. and Caridakis, G. (2019). 'A Survey of Augmented Reality Applications in Cultural Heritage'. International Journal of Computational Methods in Heritage Science. 3, pp. 118-147.

Ardissono, L., Kuflik, T., and Petrelli, D. (2012) 'Personalization in cultural heritage: the road travelled and the one ahead' User Model. User Adapt. Interact. 22(1), pp. 73–99.

Artstein, R., Traum, D., Alexander, O., Leuski, A., Jones, A., Georgila, K., Debevec, P., Swartout, W. and Maio, H. (2014) 'Time-offset interaction with a holocaust survivor' Proceedings of IUI-14, 19th International Conference on on Intelligent User Interfaces, Haifa.

Callaway, C., Stock, O., Dekoven, E. Noy, K., Citron, Y. and Dobrin, Y. (2012) 'Drama and Narrative Variation as a Means to Induce Group Conversation at the Museum' New Review of Hypermedia and Multimedia 18(1-2), pp. 37-61.

Kuflik, T., Stock, O., Zancanaro, M., Gorfinkel, A., Jbara, S., Kats, S., Sheidin, J. and Kashtan, N. (2011) 'A Visitor's Guide in an "Active Museum": Presentations, Communications, and Reflection' Journal on Computing and Cultural Heritage of the ACM, 3(3).

Kuflik, T., Wecker, A., Lanir, J. and Stock, O. (2015) 'An Integrative Framework for Extending the Boundaries of the Museum Visit Experience: Linking the Pre, During and Post Visit Phases' Journal of Information Technology and Tourism 15(1), pp. 17-47.

Stock, O., Guerini, M. and Pianesi, F. (2016) 'Ethical dilemmas for adaptive persuasion systems' Proceedings of AAAI-2016, Phoenix.

Stock, O., Zancanaro, M., Busetta, P., Callaway, C., Krüger, A., Kruppa, M., Kuflik, T., Not, E. and Rocchi, C. (2007) 'Adaptive, Intelligent Presentation of Information for the Museum Visitor in PEACH' User Modelling and User-Adapted Interaction, 17(3), pp. 257-304.

Van Camp, M. and Van den Bosch, A. (2012). 'The socialist network' Decision Support Systems. 53, pp. 761-769.

Zancanaro, M., Stock, O., Eisikovits, Z., Koren, C. and Weiss, P. L. (2012) 'Co-narrating a conflict: An interactive tabletop to facilitate attitudinal shifts' Transactions on Computer-Human Interaction (TOCHI) of the Association for Computational Machinery, 19(3), pp. 1-30.

Part V
Data, Algorithm, and Fairness

The Attention Economy and the Impact of Artificial Intelligence

Ricardo Baeza-Yates and Usama M. Fayyad

Abstract The growing ubiquity of the Internet and the information overload created a new economy at the end of the twentieth century: the economy of attention. While difficult to size, we know that it exceeds proxies such as the global online advertising market which is now over \$300 billion with a reach of 60% of the world population. A discussion of the attention economy naturally leads to the data economy and collecting data from large-scale interactions with consumers. We discuss the impact of AI in this setting, particularly of biased data, unfair algorithms, and a user-machine feedback loop tainted by digital manipulation and the cognitive biases of users. The impact includes loss of privacy, unfair digital markets, and many ethical implications that affect society as a whole. The goal is to outline that a new science for understanding, valuing, and responsibly navigating and benefiting from attention and data is much needed.

1 Introduction

We hear frequently of the information overload and its stressful impact on humanity. The growth of networks, digital communications means (most prominently email and now chat), and the plethora of information sources where access is instant and plentiful has resulted in creating a situation where we are often faced with poor, or at best questionable, quality information. This information is often misleading and potentially harmful. But this phenomenon is not limited to those seeking information or connection; the passive entertainment channels are so plentiful that even under-standing what is available is a challenge for normal humans, parents, children, educators, and professionals in general.

With hundreds of billions of web pages available (Google 2021a, b), how do you keep up? Eric Schmidt, then CEO of Google, was famously quoted as saying that humanity produced more information in 2011 alone than it did in the entire history of

R. Baeza-Yates (✉) · U. M. Fayyad
Institute for Experiential AI, Northeastern University, San Francisco, CA, USA
e-mail: rbaeza@acm.org; fayyad@acm.org

© The Author(s) 2022   123
H. Werthner et al. (eds.), *Perspectives on Digital Humanism*,
https://doi.org/10.1007/978-3-030-86144-5_18

civilization, to wit: "There were 5 Exabytes of information created between the dawn of civilization through 2003, but that much information is now created every 2 days" (Huffington Post 2011).

While the accuracy of this information was questionable, it proved to be prophetic of a much more dramatic future that became manifest in the next few years. An article in Forbes (Marr 2018) claimed: "There are 2.5 quintillion bytes of data created each day at our current pace, but that pace is only accelerating with the growth of the Internet of Things (IoT). Over the last two years alone 90% of the data in the world was generated."

A quintillion is 10^{18} bytes, hence 2.5 Exabytes per day, which about matches the quantity referenced by Eric Schmidt. Whence today, we far exceed the claims made in 2011. Disregarding the fact that most of this "data" should not be counted as data (e.g., copying a video to a distribution point does not constitute data creation in the authors' opinion), it is unquestionable that recorded information is being generated faster than any time in the history of civilization. Our ability to consume this information as humans is extremely limited and getting more challenged by the day.

The real problem is not consuming all this information, as most of it is likely to be of little value. The [now chronic] problem is one of focusing the attention on the right information. We can see this problem even in the world of science or academic publications. The ability to keep up with the quantity of publications in any field has become an impossible task for any particular researcher. Yet the growth continues. Finding the value in the right context is now much harder. Reliance on curated sources—such as journals and conferences—is no longer sufficient. Authors have many more outlets to publish papers, including sharing openly on the Web. While this appears to be a great enabler, it creates a serious problem in determining what is "trusted information." The problem is compounded by the fact that anyone can write a "news-like" article and cite these various sources. Whether real or "fake" news, these articles have a welcoming environment in social media that enables rapid spread across the Web.

So how do we deal with this growing volume of available information? While search engines have been a reasonable approach to cope with casual usage, we are far from having truly powerful search technology. Understanding the meaning and context of a short query is difficult. Search engines that have a good understanding of content and are not as reliant on statistical keyword search are not yet practical. As we explain in the next section, there are many areas of the new "attention economy" that still need new solutions that are practical. We believe that all the demand points to the need for search engines that understand semantics, context, intent, and structure of the domain being searched: from health records, to legal documents, to scientific papers, to drug discovery, to even understanding the reliability of the sources. Finally, with social media driving crowdsourcing of user-generated content, a growing importance is played by monitoring and reliably assessing the type of content being shared and whether it is appropriate by social, legal, privacy, and safety criteria and policy. These are all wide open areas needing new research.

2 The Attention Economy

Since the scarce resource is attention, an economy around getting to people's attention has rapidly grown. One only needs to examine the growth of digital advertising over the last decade or two to see economic evidence of the market value of attention.

Digital advertising on the Internet has grown from an estimated $37B market in 2007 to being well over $360B market in 2020 (Statista 2021) as shown in Fig. 1 trending the growth rate and breakdown of online advertising into the three major categories: search advertising, display advertising (graphical ads), and classified listings.

This is a good proxy as it actually underestimates the true value since many advertising markets are not yet matured. However, it is an indication of the growth of time spent by audiences online. The majority of the ad spend is on advertising on mobile devices as this medium drove most of the online growth in the last 7 years (e.g., US ad numbers (Forbes 2019)).

We use web content consumption as a proxy. The problems are actually deeper in areas that have not quite been valued economically yet, but are playing a bigger and more fundamental role in our digital lives, both as individuals and organizations/companies.

Underlying the growth of the attention economy is the *data economy* and what that data is worth to organizations and individuals. This problem is different from content on the Web and is where we have yet to see most of the growth and interesting innovations. Advertising and "hacked data" on the dark web provide underestimates of this economy. From storing the data, to maintaining it, to creating

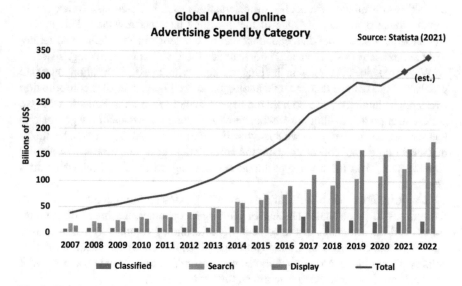

Fig. 1 Global annual online advertising spend by category (Source: Statista 2021)

cloud services to enable secure access and management, this is likely to become the big economy of the future.

Many attempts at sizing the data economy have faced many challenges. Li (2020) found that data use enables firms to derive firm-specific knowledge, which can be measured by their organizational capital, the accumulated information, or know-how of the firm (Prescott and Visscher 1980); the more the data, the greater the potential derived firm-specific knowledge. They estimated the organizational capital for each of top seven global online platform companies, Apple, Amazon, Microsoft, Google, Facebook, Alibaba, and Tencent, and compared their combined organizational capital with the global data flow during the same period of time. This provides evidence that large online platform companies have been aggressively investing capital in order to tap the economic opportunities afforded by explosive global data growth, which leads to some estimates of the size of the data economy.

No accepted methodology to measure the value of the market for data exists currently. Apart from the significant conceptual challenges, the biggest hurdle is the dearth of market prices from exchange; most data are collected for firms' own use. Firms do not release information relating to transactions in data, such as the private exchange or the sharing of data that occurs among China's big tech companies. Li et al. (2019) examine Amazon's data-driven business model from the perspectives of data flow, value creation for consumers, value creation for third parties, and monetization. Amazon uses collected consumer data and, through recommendations, insights, and third-party services, creates new products and services for its users, for example, personalized subscription services and recommendations for all products and services sold on its marketplace. "In 2019, Amazon had more than 100 million Prime subscribers in the U.S., a half of U.S. households, with the revenue from the annual membership fees estimated at over US$9 billion" (Tzuo and Weisert 2018). By taking advantage of insights on interactions and transactions, Amazon can capture much of the social value of the data they have accumulated. This is why any attempt at sizing the data economy can be challenging: the plethora of refinement and reuse forms a huge and poorly understood, yet rapidly growing, space.

What role does AI play in helping us deal with these information overload problems? We believe that since technology was the big enabler of the information overload, technology will also hold the keys to attempt to tame it. Because what we seek is an approach to filter information and focus attention, we naturally gravitate to approaches that are based on what is typically referred to as AI. The reason for this is that generating and collecting data and information does not necessarily require "intelligence." But the inverse problem: determining meaning and relevance does require it.

In order to filter, personalize, and meaningfully extract information or "reduce" it, there needs to be an understanding of the target consumer, the intent of the consumer, and the semantics of the content. Such technologies typically require either the design and construction of intelligent algorithms that code domain knowledge or employing machine learning to infer relevance from positive, negative, and relevant feedback data. We discuss this approach further in Sect. 3.

The main theme is that AI is necessary and generally unavoidable in solving this inverse problem. This brings in complications in terms of an ability to correctly address the problem, to effectively gather feedback efficiently, and, assuming the algorithms work, to issues of algorithmic bias and fairness that we shall discuss in Sect. 4.

3 The User-Machine Feedback Loop

Much of the attention economy is constructed upon the interaction feedback loop of users and systems. The components of such a loop are discussed here, and in the next section we detail the biases that poison them.

The overall setup of this problem as an AI problem is totally analogous to a search engine operation. Web search requires four major activities:

1. Crawling: web search engines employ an understanding of websites and content to decide what information to crawl, how frequently, and where to look for key information.
2. Content modeling: requires modeling concepts and equivalences, stemming, canonization (e.g., understanding what phrases are equivalent in semantics), and reducing documents to a normalized representation (e.g., bag of words, etc.).
3. Indexing and retrieval: figuring out how to look up matches and how to rank results.
4. Relevance feedback: utilizing machine learning (MLR) to optimize the matching and the ranking based on user feedback: either directly or by leveraging information like click-through rates, etc.

Each of these steps above requires some form of AI. The problem of capturing relevance and taming the information overload in general requires us to solve the equivalent components regardless of the domain of application: be it accessing a corpus of scientific applications, tracking social media content, determining what entertainment content is interesting, or retrieving and organizing healthcare information.

We note that this framework has the human-in-the-loop component for capturing the scarce resource accurately: is this content interesting for the target task? This natural loop is an example of leveraging AI either through direct feedback or through the construction of reliably labeled training data. The next three elements need serious consideration as we consider approaches to "human-in-the-loop" solutions.

3.1 Users and Personal Data

The first key element is the digital identity as a user. A digital identity can range from anonymous to a real personal identity. However, in practice, you are never completely anonymous as you can be associated with an IP address, a device identifier, or/and a browser cookie. If you are authenticated in a system, you will have more identification data associated with you, but part of it might be fake by choice.

The second element is data protection, that is, how secure is your personal data stored by the system that is using it. In many countries, this is regulated by law. The most well-known case is the General Data Protection Regulation (European Union 2016) of the European Union. This regulation includes when user consent is needed, proportionality principles for the amount of data that can be requested, how long it can be stored, and how it can be processed and transferred.

The third element is privacy, the ability of an individual or group to seclude information about themselves. For this you need at least data protection but if you can also conceal your digital identity, even better. Privacy is a shared attribute in the sense that does not depend on just the individual but the network of relations that a person has. It does not matter how private you are if the people that you know share more information about you. For this reason, some people even argue that privacy is collective (Véliz 2021) although in the United Nations Declaration is an individual human right (United Nations 1948).

3.2 Algorithms

Any software system is built on top of many algorithms. Some key components are the algorithms used to gather and store data about the environment and the users, profile the users to personalize and/or nudge their experience, and monetize their interaction. Tracking and profiling users can be done at an individual level or can be coarser, such as assigning a particular persona to each user. A "persona" is a construction of a realistic exemplar or actual specific user that represents a segment of users having certain preferences or characteristics. Apple, for example, uses differential privacy to protect individual privacy, and in the next release of its mobile operating system, iOS 14, users will decide if they can be tracked or not (O'Flaherty 2021). At the same time, Google is planning to move away from cookies by using FLoCs or Federated Learning of Cohorts (Google 2021b). Nudging implies manipulating the behavior of the user, from suggesting where to start reading to where we place elements that we want the user to click.

3.3 Digital Information Markets

The attention economy has created a specific data economy that refers to tracking and profiling users (as discussed above). For this reason, although talking about television, Serra and Schoolman in 1973 said that "It is the consumer who is consumed, you are the product of T.V." So, we are the product and the data economy is at the center of digital information markets. Digital information markets are platforms/environments that have different types of incentives that form a market. They include social networks, ecommerce, search engines, media streaming, etc. One characteristic of these markets is that they have popular as well as a long tail of items (e.g., products or movies). Users also differ on engagement, creating another long tail of interaction and hence tracked data. Most digital markets optimize the short-term revenue and hence are not necessarily optimal. All these characteristics shape the system's feedback loop.

4 Biases

In this section, we cover many relevant biases that exist in software systems, particularly ML-based systems. Bias is a systematic deviation regarding a reference value, so in principle the word is neutral. However, we usually think about bias in a negative way because in the news only negative biases are covered (gender, race, etc.). We can distinguish between statistical bias, product of a measurement; cultural or social bias; and cognitive bias that are particular to every person. We organize them by source, data and algorithms, including the interaction of users with the system.

4.1 Data

This is the main source of bias as data may encode many biases. In Table 1, we show examples of different types of generic data sets crossed with key examples of social bias that might be present, where Econ represents wealth-based discrimination. However, biases might be subtle and not known a priori. They can be explicit or implicit, direct or indirect. In addition, sometimes it is not clear what should be the right reference value or distribution (e.g., gender or age in a given profession).

One important type of data that encodes many biases is text. In addition to gender or racial bias, text can encode many cultural biases. This can even be seen when it is used to train word embeddings, large dimensional spaces where every word is encoded by a vector. There are examples of gender bias (Caliskan et al. 2017),

Table 1 Example of biases found on different types of data sets

Data set	Gender	Race	Age	Geo	Econ
Faces	✓	✓	✓	✓	✓
News	✓	✓	✓	✓	✓
Resumes	✓	✓	✓	✓	✓
Immigration	✓	✓		✓	✓
Criminality		✓		✓	✓
Recidivism		✓		✓	✓

race bias (Larson et al. 2016), religious bias (Abid et al. 2021), etc., and their impact has many ramifications (Bender et al. 2021).

There can be biases also in how we select data. The first one is the sample size. If the sample is too small, we bias the information (say events) in the sample to the most frequent ones. This is very important in Internet data as the probability of an event, say a click, is very small and the standard sample size formula will underestimate the real value. Hence, we need to use adapted formulas to avoid discarding events that we truly want to measure (Baeza-Yates 2015). In addition, in the Internet, event distributions usually follow a power law with a very long constant tail and hence are very skewed. For this type of distribution, the standard sampling algorithm does not generate the same distribution in the sample, particularly in the long tail. Hence, it is important to use a stratified sampling method in the event of interest to capture the correct distribution.

4.2 Algorithms

Bias in algorithms is more complicated. A classic example is recommending labels in a platform that gathers labels from users. The availability of recommendations will incentivize users to put less labels. If the system recommends labels without any seed label, there will be no more new labels, and hence the algorithm cannot learn anything new. In other words, the algorithm itself kills the label folksonomy (ironically helping people reduce data coming from them!).

In other cases, the function that the algorithm optimizes, perhaps designed with best intentions, produces new or amplifies existing bias. Examples include race bias amplification in bail prediction (Kleinberg et al. 2018) or delivery time bias in food delivery (Forbes 2021). Part of the problem here is that many times the focus of the designers is on increasing revenue without thinking about the indirect impact of what is being optimized. Moreover, recently, the first indication that programmers can transfer cognitive biases to code was published (Johansen et al. 2020), an example of more subtle indirect bias.

The largest source of algorithmic bias is in the interaction with the user. First, we have exposure or presentation bias. Users can only interact with what is presented to them and that is decided by the system. Hence, if the system uses interaction data for, say, personalization, the system is partly shaping that data. Because of the attention economy, the amount of interaction of users also follows a power law, creating an engagement or participation bias (Nielsen 2016; Baeza-Yates and Saez-Trumper 2015).

One of the main effects of the interaction feedback loop is popularity bias. That is, popular items get more attention than they deserve with respect to items in the long tail. Other effects depend on cognitive biases of the users such as how they look at the screen, how much they are influenced by the information read, or how often they click or move the mouse. This creates position bias and, in the case of search engines, ranking bias (Baeza-Yates 2018). That is, top ranking positions get more clicks only because they are at the top. To counteract this bias, search engines debias clicks to avoid fooling themselves. In addition, ratings from other users create social pressure bias.

5   Societal Impact

There are several areas of impact on society which are summarized by:

1. **How the data economy creates a loss of privacy.** A loss of privacy that many people are not aware of as they normally do not read the terms of usage. Shoshana Zuboff calls this *surveillance capitalism* (Zuboff 2019), or surveillance economy, to distinguish it from government surveillance, as it is mainly carried out by large Internet multinationals. Carissa Véliz (2021) argues that "whoever has the most personal data will dominate society." Hence, privacy is a form of power: if companies control it, the wealthy will rule; while if governments control it, dictatorships will rule. The conclusion is that society can only be free if people keep their power, that is, their data. She goes on augmenting that data is a toxic substance that is poisoning our lives and that economic systems based on the violation of human rights are unacceptable, not only because of ethical concerns but because the "surveillance economy threatens freedom, equality, democracy, autonomy, creativity, and intimacy."

2. **Digital manipulation of people.** This goes beyond the digital nudging explained earlier, say to entice you to click an ad without your noticing it. The main example is social media and fake news. This new social age is ruled by what Sinan Aral calls *The Hype Machine* (Aral 2020), a machine that has disrupted elections, stock markets, and, with the COVID-19 pandemic, also our health. There are many examples of country-wide manipulation from governments such as in Brazil, Myanmar, and the Philippines or from companies such as Cambridge Analytica using Facebook data which affected the 2016 US presidential election. Harari is much more pessimistic as some fake news lasts forever and "as a species, humans prefer power to truth" (Harari 2018). The future danger for him is the combination of AI with neuroscience, the direct manipulation of the brain. In all of the above, AI is the key component to predict which person is more prone to a given nudging and how to perform it.

3. **Unfair digital markets** (monopolistic behavior and the biased feedback loop previously mentioned). During 2020, the US government started antitrust cases against Facebook (US FTC 2020) and Google (US DoJ 2020). However, the popularity bias also discriminates against the users and items in the tail, creating unfairness and also instability. Better knowledge of the digital market should imply optimal long-term revenue and healthier digital markets, but the current recommendation systems are optimized for short-term revenue. In some sense, the system is inside a large echo chamber that is the union of all the echo chambers of its users (Baeza-Yates 2020).

4. **Ethical implications of all the above**. That includes discrimination (e.g., persons, products, businesses), phrenology (e.g., prediction of personality traits based in facial biometrics (Wang and Kosinski 2018)), data used without consent (e.g., faces scrapped from the Web used to train facial recognition systems (Raji and Fried 2020)), etc.

There are also impacts in specific domains such as government, education, health, justice, etc. Just to exemplify one of them, we analyze the impact on healthcare. While machine-driven and digital health solutions have created a huge opportunity for advancing healthcare, the information overload has also affected our ability to leverage the richness of all the new options: genome sequencing, drug discovery, and drug design vs. the problem of understanding and tracking what this data means to our lives. How the healthcare industry is having trouble dealing with all the data is manifested in these examples:

- Lack of standards created data swamps in electronic health records (EHRs)
- Lack of ability to leverage data for population healthcare studies and identify panels on demand
- Lack of ability to leverage collected data to study drug and treatment impacts systematically
- Inability of healthcare providers to stay on top of the data for an individual
- Lack of individuals in ability to stay on top of data about themselves

6 Conclusions

As the tide of information overload rises, we believe that this makes the traditional long tail problem even harder. The reason for this is that the information coming from the head of the tail, the most common sources of new data or information, is not growing as fast as information coming from the less popular sources that are now enabled to produce more growth. This exacerbates the long tail distribution and makes information retrieval, search, and focus of attention much more difficult.

As a final concluding thought, one may wonder if the only way out is through use of AI algorithms. The answer is a qualified yes. There may be better solutions through design, through proper content tagging and classification and modeling as content is generated. But the reality is that for the majority of the time, we are living in a world where we have to react to new content, new threats, new algorithms, and new discovered biases. As such, we will always be left with the need for a scalable approach that has to solve the "inverse design" problem—inferring from observations what is likely happening. This seems to drive "understanding," especially semantic modeling, to the forefront. And this seems to drive us to look for algorithms to solve such inference problems, and whence AI.

There are other related issues that we did not cover. Those include cybersecurity and the growing dark web economy, as well as other emerging technologies that create synergy with AI. The same for the future impact of the proposed regulation for AI in the European Union that was just published (EU 2021).

References

Abid, Abubakar; Farooqi, Maheen; Zou, James. (2021) Persistent Anti-Muslim Bias in Large Language Models. https://arxiv.org/pdf/2101.05783v1.pdf

Aral, Sinan. (2020) The Hype Machine, Currency Press.

Baeza-Yates, Ricardo and Saez-Trumper, Diego. (2015) Wisdom of the crowd or wisdom of a few? An analysis of users' content generation. In *Proceedings of the 26th ACM Conference on Hypertext and Social Media* (Guzelyurt, TRNC, Cyprus, Sept. 1–4). ACM Press, New York, 69–74.

Baeza-Yates, Ricardo. (2015) Incremental sampling of query logs. In *Proceedings of the 38th ACM SIGIR Conference* (Santiago, Chile, Aug. 9–13). ACM Press, New York, 1093–1096.

Baeza-Yates, Ricardo. (2018) Bias on the Web. Communications of ACM 61(6), 54-61.

Baeza-Yates, Ricardo. (2020) Bias in Search and Recommender Systems. ACM RecSys 2020, Rio de Janeiro. https://www.youtube.com/watch?v=8zetbdx4_08

Bender, Emily M.; Gebru, Timnit; McMillan-Major, Angelina; Mitchell, Margaret. (2021) On the Dangers of Stochastic Parrots: Can Language Models Be Too Big?. ACM FAccT 2021. https://faculty.washington.edu/ebender/papers/Stochastic_Parrots.pdf

Caliskan, Aylin; Bryson, Joanna J. and Narayanan, Arvind. (2017) Semantics derived automatically from language corpora contain human-like biases. Science 356, 6334, 183–186.

European Union. (2016) General Data Protection Regulation 2016/679.

European Union. (2021) Proposed Regulation for an European Approach to AI. https://digital-strategy.ec.europa.eu/en/library/proposal-regulation-european-approach-artificial-intelligence

Forbes. (2019) https://www.forbes.com/sites/greatspeculations/2019/06/11/how-has-the-u-s-online-advertising-market-grown-and-whats-the-forecast-over-the-next-5-years/

Forbes. (2021) https://www.forbes.com/sites/jonathankeane/2021/01/05/italian-court-finds-deliveroo-rating-algorithm-was-unfair-to-riders/

Google. (2021a) How Search Works. https://www.google.com/search/howsearchworks/crawling-indexing/

Google. (2021b) Federated Learning of Cohorts. https://github.com/WICG/floc

Harari, Yuval Noah. (2018) 21 Lessons for the 21st Century. Spiegel & Grau.

Huffington Post, (2011) Google CEO Eric Schmidt: 'People Aren't Ready for The Technology Revolution'. https://www.huffpost.com/entry/google-ceo-eric-schmidt-p_n_671513

Johansen, Johanna; Pedersen, Tore; Johansen, Christian. (2020) Studying the Transfer of Biases from Programmer to Programs. arXiv, https://export.arxiv.org/pdf/2005.08231

Kleinberg, Jon; Lakkaraju, Himabindu; Leskovec, Jure; Ludwig, Jens and Mullainathan, Sendhil. (2018) Human Decisions and Machine Predictions, *The Quarterly Journal of Economics*, Oxford University Press, vol. 133(1), 237-293.

Larson, Jeff; Mattu, Surya; Kirchner, Lauren; Angwin, Julia. (2016) How We Analyzed the COMPAS Recidivism Algorithm. https://www.propublica.org/article/how-we-analyzed-the-compas-recidivism-algorithm

Li, Wendy C.Y.; Nirei, Makoto; Yamana, Kazufumi. (2019) Value of Data: There's No Such Thing as a Free Lunch in the Digital Economy, U.S. Bureau of Economic Analysis, working paper, https://www.bea.gov/system/files/papers/20190220ValueofDataLiNireiYamanaforBEAworkingpaper.pdf

Li, Wendy C.Y. (2020) Online Platforms' Creative "Disruption" in Organizational Capital – the Accumulated Information of the Firm, U.S. Bureau of Economic Analysis working paper.

Marr, B. (2018) How Much Data Do We Create Every Day? Forbes. https://www.forbes.com/sites/bernardmarr/2018/05/21/how-much-data-do-we-create-every-day-the-mind-blowing-stats-everyone-should-read/

Nielsen, Jakob. (2016) The 90-9-1 Rule for Participation Inequality in Social Media and Online Communities. https://www.nngroup.com/articles/participation-inequality/

O'Flaherty, Kate. (2021) Apple's Stunning iOS 14 Privacy Move. Forbes. https://www.forbes.com/sites/kateoflahertyuk/2021/01/31/apples-stunning-ios-14-privacy-move-a-game-changer-for-all-iphone-users/

Prescott, E. and Visscher, M. (1980) *Journal of Political Economy*, vol. 88, issue 3, 446-61.

Raji, Inioluwa Deborah and Fried, Genevieve. (2020) About Face: A Survey of Facial Recognition Evaluation, AAAI 2020 Workshop on AI Evaluation.

Statista (2021) https://www.statista.com/statistics/276671/global-internet-advertising-expenditure-by-type/

Tzuo, Tien and Weisert, Gabe. (2018) Subscribed: Why the Subscription Model Will be Your Company's Future – And What to Do about It, Publisher: Portfolio/Penguin.

United Nations. (1948) Declaration of Human Rights, Article 12.

United States Department of Justice (2020). https://www.justice.gov/opa/pr/justice-department-sues-monopolist-google-violating-antitrust-laws

United States Federal Trade Commission (2020. https://www.ftc.gov/news-events/press-releases/2020/12/ftc-sues-facebook-illegal-monopolization

Véliz, Carissa. (2021) Privacy is Power. Bantam Press, Second edition.

Wang, Yilun; Kosinski, Michal. (2018) Deep neural networks are more accurate than humans at detecting sexual orientation from facial images. J. of Personality and Social Psychology;114 (2):246-257.

Zuboff, Shoshana. (2019) The Age of Surveillance Capitalism. Public Affairs.

Did You Find It on the Internet? Ethical Complexities of Search Engine Rankings

Cansu Canca

Abstract Search engines play a crucial role in our access to information. Their search ranking can amplify certain information while making others virtually invisible. Ethical issues arise regarding the criteria that the ranking is based on, the structure of the resulting ranking, and its implications. Critics often put forth a collection of commonly held values and principles, arguing that these provide the needed guidance for ethical search engines. However, these values and principles are often in tension with one another and lead us to incompatible criteria and results, as I show in this short chapter. We need a more rigorous public debate that goes beyond principles and engages with necessary value trade-offs.

1 Introduction

In our digitalized life, search engines function as the gatekeepers and the main interface for information. Ethical aspects of search engines are discussed at length in the academic discourse. One such aspect is the *search engine bias*, an issue that encompasses ethical concerns about search engine rankings being value-laden, favoring certain results, and using non-objective criteria (Tavani 2020). The academic ethics debate has not (yet) converged to a widely accepted resolution of this complex issue. Meanwhile, the mainstream public debate mostly ignored the hard ethical trade-offs, opting instead for a collection of commonly held principles and values such as accuracy, equality, efficiency, and democracy. In this short chapter, I explain why this approach leads to unresolvable conflicts and thus ultimately a dead end.

This chapter builds on the *Mapping* workshop (https://aiethicslab.com/the-mapping/) I designed and developed together with Laura Haaber Ihle at AI Ethics Lab. I thank Laura also for her insights and valuable feedback on this chapter.

The original version of this chapter was revised with an update in the affiliation of the author Cansu Canca. A correction to this chapter can be found at https://doi.org/10.1007/978-3-030-86144-5_47

C. Canca (✉)
AI Ethics Lab, Cambridge, MA, USA
e-mail: cansu@aiethicslab.com

2   Value of and Value Within Search Engines

Search engines are invaluable. Without them, it is impossible to navigate the massive amounts of information available in the digital world. They are our main mediator of information (CoE 2018). Every day over 7 billion searches are conducted on Google alone,[1] accounting for over 85% of all searches worldwide.[2] In addition to that, various searches are conducted on specialized search engines such as Amazon and YouTube. Even academic and scientific research relies on Google Scholar, PubMed, JSTOR, and similar specialized search engines—meaning not only we rely on search engines to access information but we also rely on them while creating further knowledge.

We defer to search engines' ranking of information so much that most people do not even check the second page of the search results.[3] On Google, 28.5% of users click the first result, while 2.5% click on the tenth result, and fewer than 1% click results on the second page.[4] This means that the ranking has a great effect on what people see and learn.

Search engine ranking is never value-neutral or free from human interference. It is optimized to reach a goal. Even if we can agree that this goal is *relevance* to the user, defining what is relevant involves guesswork and value judgments. By definition, in most search queries, users do not know what result they are looking for. On top of that, the search engine has to guess from the search keywords what *kind* of information the user is interested in. For example, a user searching for "corona vaccination" could be looking for vaccine options, vaccine safety information, anti-vaccine opinions, vaccination rates, or celebrities who are vaccinated, and they might be looking for these on a global or local scale. More importantly, they might be equally satisfied with well-explained vaccine safety information or anti-vaccine opinions since they might not have prior reasons to differentiate these two opposing results. Here, the value judgment comes into play in designing the system. Should the system first show vaccine safety information to ensure that the user is well-informed or the anti-vaccine opinions since they are often more intriguing and engaging? Should the system make results depending on user profiles (e.g., being scientifically or conspiracy oriented)? Should it sort the results by click rates globally or locally, by personal preferences of the user, by accuracy of the information, by balancing opinions, or by public safety? Deciding which ranking to present embeds a value judgment into the search engine. And this decision cannot fully rely on evaluating user satisfaction about a search query, because the user does not know the full range

---

[1] https://www.internetlivestats.com/google-search-statistics/

[2] https://www.statista.com/topics/1710/search-engine-usage/-dossierSummary

[3] https://www.forbes.com/sites/forbesagencycouncil/2017/10/30/the-value-of-search-results-rankings/

[4] https://www.sistrix.com/blog/why-almost-everything-you-knew-about-google-ctr-is-no-longer-valid/

of information they could have been shown. Moreover, user satisfaction might still lead to unethical outcomes.

3 Ethical Importance of Search Engine Rankings

Imagine basing your decision whether to get vaccinated on false information because that is what came up on your search (Ghezzi et al. 2020; Johnson et al. 2020). Imagine deciding whom to vote for based on conspiracy theories (Epstein and Robertson 2015). Imagine having your perception of other races and genders pushed to negative extremes because that is the stereotype you are presented with online (Noble 2018). Imagine searching for a job but not seeing any high pay and high power positions because of the search engine's knowledge or estimate of your race, gender, or socio-economic background.[5] Imagine having your CV never appear to the employers for those jobs that you easily qualify for because of search engine profiling (Deshpande et al. 2020). These are all ethical issues. They stem from value judgments embedded in search engine processing and, as a result, impacting individual autonomy, well-being, and social justice.

We base our decisions on what we know. By selecting which information to present in what order, search engines affect and shape our perception, knowledge, decision, and behavior both in the digital and physical sphere. As a result, they can manipulate or nudge individuals' decisions, interfere with their autonomy, and affect their well-being.

By sorting and ranking what is out there in the digital world, search engines also impact how benefits and burdens are distributed within the society. In a world where we search online for jobs, employees, bank credits, schools, houses, plane tickets, and products, search engines play a significant role in the distribution of opportunities and resources. When certain goods or opportunities are systematically concealed from some of us, there is an obvious problem of equality, equal opportunity and resources, and, more generally, fairness. More to the point, once certain information is not accessible to some, they often do not even know that it exists. If we cannot realize the injustice, we also cannot demand a fair treatment.

4 Do You See Female Professors?

A running example for search engine bias has been the image search results for the term "professor."[6] When searching for "professor," search engines present a set of overwhelmingly white male images. In the USA, for instance, women make up 42%

---

[5] https://www.nytimes.com/2018/09/18/business/economy/facebook-job-ads.html

[6] https://therepresentationproject.org/search-engine-bias/

of all tenured or tenure-track faculty.[7] In search engine results, the ratio of female images has been about 10–15% and only recently went up to 20–22%.[8] When searching specifically for "female professors," the image results are accompanied by unflattering suggestions: Google's first suggested term is "clipart" (Fig. 1), whereas Bing's top four suggestions include "crazy," "cartoon," and "clipart" female professors (Fig. 2).[9]

Why is this an ethical problem? Studies show that girls are more likely to choose a field of study if they have female role models (Lockwood 2006; Porter and Serra 2020). Studies also show that gender stereotypes have a negative effect on hiring women for positions that are traditionally held by men (Isaac et al. 2009; Rice and Barth 2016). By amplifying existing stereotypes, search engine results contribute to the existing gender imbalance in high-powered positions. It is reasonable to think that this gender imbalance in real life has its roots in unjustified discrimination against women in the workplace as well as discriminatory social structures, both of which do not allow female talent to climb the career ladder. The search engine bias can contribute to the perpetuation of this gender imbalance.

In fact, this is not special to image search results for "professor." Women are underrepresented across job images and especially in high-powered positions (Kay et al. 2015; Lam et al. 2018). Take one step further in this problem, and we end up with issues such as LinkedIn—a platform for professional networking—autocorrecting female names to male ones in its search function[10] and Google showing much fewer prestigious job ads to women than to men.[11]

Most mainstream reactions criticize search engine rankings from commonly held values and principles: search engines should reflect the truth and be fair; they should promote equality and be useful; they should allow users to exercise agency and prevent harm; and so on. On close inspection, however, these commonly held values and principles fail to provide guidance and may even conflict with one another. In the next paragraphs, I briefly go over three values—accuracy, equality, and agency—to show how such simple guidance is inadequate for responding to this complex problem.

Accuracy One could argue that search engines, being a platform for information, should accurately reflect the world as it is. This would imply that the image results

---

[7] https://www.aaup.org/news/data-snapshot-full-time-women-faculty-and-faculty-color.

[8] Calculated from search engine results of Google, Bing, and DuckDuckGo in May 2021. Note that these numbers fluctuate depending on various factors about the user such as their region and prior search results.

[9] In comparison, Bing's suggestions when searching for "male professor" remain within the professional realm: "black professor," "old male teacher," "black female professor," "black woman professor," "professor student," and such. However, Google fails equally badly in its suggestions for "male professor." Its first two suggestions are "stock photo" and "model."

[10] https://phys.org/news/2016-09-linkedin-gender-bias.html

[11] https://www.washingtonpost.com/news/the-intersect/wp/2015/07/06/googles-algorithm-shows-prestigious-job-ads-to-men-but-not-to-women-heres-why-that-should-worry-you/

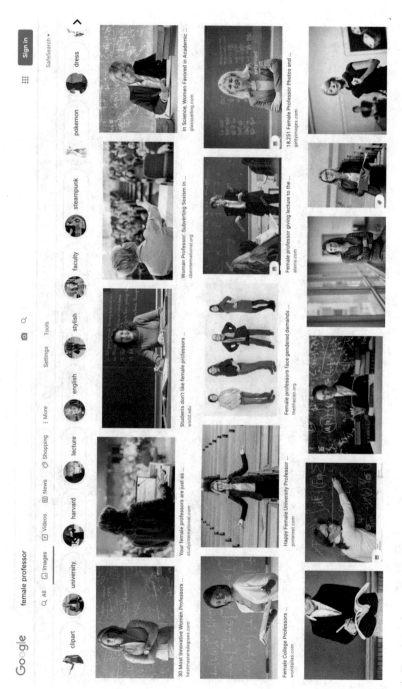

Fig. 1 Female professors and Google

Fig. 2 Female professors and Bing

should be revised and continuously updated to reflect the real-life gender ratio of professors. In contrast, the current search results portray the social perception about gender roles and prestigious jobs. Note that implementing accuracy in search results would require determining the scope of information: Should the results accurately reflect local or global ratio? And what other variables—such as race, ability, age, and socio-economic background—and their ratio should the results reflect accurately?

Equality Contesting prioritization of accuracy, one could argue that search engines should promote equality because they shape perception while presenting information, and simply showing the status quo despite the embedded inequalities would be unfair. If we interpret equality as equal representation, this would imply showing equal number of male and female professors. Implementing equal representation in search results would also require taking into account other abovementioned variables—such as race, ability, age, and socio-economic background—and equally representing all their combinations. A crucial question would then be, what would the search results look like if all possible identities are represented and would these results still be relevant, useful, or informative for the user?

Agency Contesting both accuracy and equality, one could argue that the system should prioritize user agency and choice by ranking the most clicked results at the top. This is not a strong argument. When conducting a search, users do not convey an informed and intentional choice through their various clicks. One could, however, incorporate user settings to the search engine interface to encourage user agency and provide them with a catalogue of setting options for ranking. Yet, since most people have psychological inertia and status quo bias, most users would still end up using the default (or easiest) setting—which brings us back to the initial question: What should the default ranking be?

An additional consideration must be the content of web pages that these images come from. It is not sufficient to have an ethically justifiable combination of images in the search results. It is also important that the web pages that link to these images follow a similar ethical framework. If, for example, search results show equal number of female and male images but the pages with female images dispute gender equality, this would not be a triumph of the principle of equality.

We could continue with other values such as well-being, efficiency, and democracy. They would yield even more different and conflicting ranking outcomes. Therein lies the problem. These important and commonly held values do not provide a straightforward answer. They are often in tension with each other. We all want to promote our commonly cherished and widely agreed-upon values. This is apparent from the Universal Declaration of Human Rights, from the principlism framework, and from the common threads within various sets of AI principles published around the world.[12] But simply putting forth some or all of these values and principles and

---

[12] https://aiethicslab.com/big-picture/

demanding them to be fulfilled is an unreasonable and impossible request, which do not get us very far in most cases (Canca 2019, 2020).

5 The Process and the End Product

Thus far I focused on the end product. What is the ethically justified composition for search engine results? But we also need to focus on the process: how did we end up with the current results, and what changes can or should be implemented?

Search engines use a combination of proxies for "relevance." In addition to the keyword match, this might include, for example, click rates, tags and indexing, page date, placing of the term on the website, page links, expertise of the source, and user location and settings.[13] Search is a complex task, and the way that these proxies fit together changes all the time. Google reports that in 2020 they conducted over 600,000 experiments to improve their search engine and made more than 4500 changes.[14] Since search engine companies compete in providing the most satisfactory user experience, they do not disclose their algorithms.[15] Returning to our example, when we compare image search results for "professor" in Google, Bing, and DuckDuckGo, we see that while Google prioritizes Wikipedia image as the top result, Bing and DuckDuckGo refer to news outlet and blog images for their first ten results, excluding the image from the Wikipedia page.[16]

Value judgments occur in deciding which proxies to use and how to weigh them. Ensuring that the algorithm does not fall into ethical pitfalls requires navigating existing and expected ethical issues. Going back to our example, content uploaded and tagged by users or developers is likely to carry their implicit gender biases. Therefore, it is reasonable to assume that the pool of images tagged as "professor" would be highly white-male oriented to start with. Without any intervention, this imbalance is likely to get worse when users click on the results that match their implicit bias and/or when an algorithm tries predicting user choice and, thereby, user bias.

---

[13] https://www.google.com/search/howsearchworks/algorithms/; https://www.bing.com/webmasters/help/webmasters-guidelines-30fba23a; https://dl.acm.org/doi/abs/10.1145/3361570.3361578

[14] https://www.google.com/search/howsearchworks/mission/users/

[15] Even if they did disclose their algorithms, this would likely be both extremely inefficient and ethically problematic. See Grimmelmann (2010) for a more detailed discussion of transparency in search engines.

[16] At the height of these discussions in 2019, the image in the Wikipedia entry for "professor" was switched to Toni Morrison, a black female professor, thereby making this the first image to appear on Google image search results. Currently (May 2021), the Wikipedia image has been changed to a picture of Einstein.

6 Conclusions

A comprehensive ethical approach to search engine rankings must take into account search engines' impact on individuals and society. The question is how to mitigate existing ethical issues in search engine processing and prevent amplifying them or creating new ones through user behavior and/or system structure. In doing so, values and principles can help us formulate and clarify the ethical problems at hand, but they cannot guide us to a solution. For that, we need to engage in deeper ethics analyses which provide insights to the value trade-offs and competing demands that we must navigate to implement any solutions to these problems. These ethics analyses should then feed into the public debate so that the discussion can go beyond initial arguments, reveal society's value trade-offs, and be informative for decision-makers.

References

Canca, C. (2019) 'Human Rights and AI Ethics – Why Ethics Cannot be Replaced by the UDHR', in *AI & Global Governance, United Nations University Centre for Policy Research* [online]. Available at: https://cpr.unu.edu/publications/articles/ai-global-governance-human-rights-and-ai-ethics-why-ethics-cannot-be-replaced-by-the-udhr.html (Accessed: 1 May 2021).

Canca, C. (2020) 'Operationalizing AI Ethics Principles', *Communications of the ACM*, 63(12), pp.18-21. https://doi.org/10.1145/3430368

Council of Europe (CoE) (2018) *Algorithms and Human Rights* [online]. Available at: https://rm.coe.int/ algorithms-and-human-rights-en-rev/16807956b5 (Accessed: 1 May 2021)

Deshpande, K.V., Shimei, P. and Foulds, J.R. (2020) 'Mitigating Demographic Bias in AI-based Resume Filtering', in *Adjunct Publication of the 28th ACM Conference on User Modeling, Adaptation and Personalization.* https://doi.org/10.1145/3386392.3399569

Epstein, R. and Robertson R.E. (2015) 'The search engine manipulation effect (SEME) and its possible impact on the outcomes of elections', *Proceedings of the National Academy of Sciences*, 112(33). https://doi.org/10.1073/pnas.1419828112

Grimmelmann, J. (2010) 'Some Skepticism About Search Neutrality' [online]. Available at: http://james.grimmelmann.net/essays/SearchNeutrality (Accessed: 1 May 2021)

Ghezzi, P., Bannister, P.G., Casino, G., Catalani, A., Goldman, M., Morley, J., Neunez, M., Prados-Bo, A., Smeesters, P.R., Taddeo, M., Vanzolini, T. and Floridi, L. (2020) 'Online Information of Vaccines: Information Quality, Is an Ethical Responsibility of Search Engines', *Frontiers in Medicine*, 7(400). https://doi.org/10.3389/fmed.2020.00400

Isaac, C., Lee, B. and Carnes, M. (2009) 'Interventions that affect gender bias in hiring: a systematic review', *Academic medicine: Journal of the Association of American Medical Colleges*, 84(10), pp.1440–1446. https://doi.org/10.1097/ACM.0b013e3181b6ba00

Johnson, N.F., Velásquez, N., Restrepo, N.J., Leahy, R., Gabriel, N., El Oud, S., Zheng, M., Manrique, P., Wuchty, S. and Lupu, Y. (2020) 'The online competition between pro- and anti-vaccination views', *Nature*, 582, pp.230–233. https://doi.org/10.1038/s41586-020-2281-1

Kay, M., Matuszek, C. and Munson, S.A. (2015) 'Unequal Representation and Gender Stereotypes in Image Search Results for Occupations', in *Proceedings of the 33rd Annual ACM Conference on Human Factors in Computing Systems.* https://doi.org/10.1145/2702123.2702520

Lam, O., Wojcik, S., Broderick, B. and Hughes, A. (2018) 'Gender and Jobs in Online Image Searches', *Pew Research Center* [online]. Available at: https://www.pewresearch.org/social-trends/2018/12/17/gender-and-jobs-in-online-image-searches/ (Accessed: 1 May 2021)

Lockwood, P. (2006) 'Someone Like Me can be Successful: Do College Students Need Same-Gender Role Models?', *Psychology of Women Quarterly*, 30(1), pp.36–46. https://doi.org/10.1111/j.1471-6402.2006.00260.x

Noble, S.U. (2018) *Algorithms of Oppression: How Search Engines Reinforce Racism*, New York: New York University Press.

Porter, C. and Serra, D. (2020) 'Gender Differences in the Choice of Major: The Importance of Female Role Models', *American Economic Journal: Applied Economics*, 12(3), pp.226–254. https://doi.org/10.1257/app.20180426

Rice, L. and Barth, J.M. (2016) 'Hiring Decisions: The Effect of Evaluator Gender and Gender Stereotype Characteristics on the Evaluation of Job Applicants', *Gender Issues* 33, pp.1–21. https://doi.org/10.1007/s12147-015-9143-4

Tavani, H. (2020) 'Search Engines and Ethics', *The Stanford Encyclopedia of Philosophy*, Fall 2020 Edition in Zalta, E.N. (ed.) [online]. Available at: https://plato.stanford.edu/archives/fall2020/ entries/ethics-search/ (Accessed: 1 May 2021)

Personalization, Fairness, and Post-Userism

Robin Burke

Abstract The incorporation of fairness-aware machine learning presents a challenge for creators of personalized systems, such as recommender systems found in e-commerce, social media, and elsewhere. These systems are designed and promulgated as providing services tailored to each individual user's unique needs. However, fairness may require that other objectives, possibly in conflict with personalization, also be satisfied. The theoretical framework of post-userism, which broadens the focus of design in HCI settings beyond the individual end user, provides an avenue for this integration. However, in adopting this approach, developers will need to offer new, more complex narratives of what personalized systems do and whose needs they serve.

1 Introduction

The turn toward questions of fairness in machine learning (Barocas and Selbst 2016; Dwork et al. 2012; Mitchell et al. 2021) raises some important issues for the understanding of personalized systems. Researchers studying these systems and organizations deploying them present a common narrative highlighting the benefits of personalization for the end users for whose experience such systems are optimized. This narrative in turn shapes users' expectations and their *folk theories* (working understandings) about the functionality and affordances of personalized systems (DeVito et al. 2017). Fairness requires a different kind of analysis. Rather than focusing on the individual, fairness is understood in terms of distribution: how is harm or benefit from a system distributed over different individuals and/or different classes of individuals? These distributional concerns take the focus at least partly away from the end user, and thus the implementation of fairness concerns in personalized systems requires a re-thinking of fundamental questions about what personalized systems are for and what claims should be made about them.

R. Burke (✉)
University of Colorado, Boulder, CO, USA
e-mail: robin.burke@colorado.edu

A personalized system tailors its presentation to an evolving understanding of what a user seems to want or need in the present moment. What are we to do when the thing that the user wants contributes to unfairness? To make this question more concrete, let us consider a particular class of personalized system, recommender systems, which filter and prioritize information and items for users in a personalized manner. Common examples include applications embedded in social media, streaming audio and video, news, job hunting, and e-commerce sites. Consider a recommender system embedded in an employment-oriented site such as XING or LinkedIn that supports human resource professionals in locating job candidates for positions. It may be that the user (the recruiter in this case) by virtue of their interactions builds a profile that disadvantages applicants in a protected category: women, for example. The recruiter's recommendation lists are disproportionately filled with male candidates, and they might not even realize the sequence of events bringing this about. Worse yet, the prevalence of male candidates will likely cause the recruiter's interactions to be primarily with male applicants, generating even more evidence (from the system's perspective) of their interest in these candidates over female or non-binary ones. The system thereby becomes what O'Neil (2016) calls a "weapon of math destruction," a computationally governed feedback cycle generating ever societally worse outcomes.

From the standpoint of pure personalization, this outcome might appear to be a success story: the user tells the system what they want and the system delivers, getting "better" at doing so over time. However, it should be clear that this is a highly undesirable outcome. A system that perpetuates systemic unfairness, even if only by passing on biases from its input to its output, by definition becomes itself part of the oppressive system (Kendi 2019). If we are interested in the beneficence of personalization, we cannot ignore this risk, and therefore we are called up on to re-consider the concept of personalization itself. The concept of "post-userism" as articulated by Baumer and Brubaker (2017) is a theoretical approach that calls into question the user focus that has dominated study in human-computer interaction since the field's inception and raises the possibility that, in understanding and evaluating computing systems, a larger and more complex framing may be essential. Our somewhat dystopian, although hardly unrealistic, personalized recruiting system suggests precisely that we need to look beyond the end user to understand how to build recommender systems free from such harmful effects.

2   De-centering the User

To de-center the user in recommendation is to consider the possibility of additional stakeholders whose objectives and goals should be integrated into the generation of recommendations. The concept of the stakeholder emerges in the management literature in the mid-1960s (as a contrast to the shareholder) defined by some authors as "any groups or individuals that can affect, or are affected by, the firm's objectives" (Freeman 2010). A multiplicity of non-user considerations, especially business objectives, have entered into practical recommender systems designs from their first deployments. However, businesses that employ such recommendation objectives have generally been very reluctant to identify anything other than user benefit as a driver for their technical decisions. An explicit consideration of multistakeholder objectives and one that specifically incorporates fairness is much more recent (Abdollahpouri et al. 2020).

What might a post-userist take on personalization look like? We examine multistakeholder considerations that can help answer this question.

First, consider the recruiter-oriented recommender system outlined above. The challenge here is to achieve *provider-side fairness* (R. Burke 2017): fair representation across those providing items to be recommended, in this case the job applicants. So, a desired system design is one in which there are limits to the degree of personalization that can be performed, even for a single user.

The system would need to ensure that each recommendation list has at least a minimal degree of diversity across different protected group categories. One would expect that an end user/recruiter would need to know going in (and might even require as a matter of law or company policy) that the recommender is enforcing such fairness constraints.

A more relaxed version of this provider-side constraint might appear in a consumer taste domain, such as the recommendation of music tracks in a streaming music service. The organization might have the goal of fair exposure of artists across different demographic groups or across different popularity categories (Mehrotra et al. 2018). List-wise guarantees might not be important, as there may be some users with very narrow musical tastes and others who are more ecumenical. As long as the goal of equalizing exposure is met, the precise distribution of that exposure over the user population might be unimportant. In this case, it may be desirable to differentiate between types of users for the purposes of fair exposure as in Liu et al. (2019) or to more precisely target the types of diversity of interest to individual users (Sonboli et al. 2020). A system that worked in this manner might need to inform users that non-personalization objectives such as fairness are operative in the recommendations it produces.

An important distinction between the cases above is that music tracks are non-rivalrous goods: a music track can be played for any number of users, and its utility is unaffected by the number of recommendations or their occurrence in time. A job candidate is different. A highly qualified candidate may be present in the job market for a very limited period of time. A recruiter who is recommended such a candidate as soon as their resume appears in the system gets greater utility from the recommendation than does a user who gets it later. A situation in which a highly qualified candidate appears only to a limited number of recruiters is more valuable to them than a situation in which their recommendations are shared with a larger group. One could imagine that the recruiter-users would be rightly concerned that the inclusion of multistakeholder considerations in recommendation could put them at a disadvantage relative to the purely personalized status quo. I say "might" here because the close study of multistakeholder recommendation is sufficiently new that it is unclear what the interactions are between recommendation quality as experienced by users and the fairness properties of the associated results. Preliminary results in some applications indicate that simultaneous improvements in both dimensions may be possible (Mehrotra et al. 2018). Where there is a tradeoff between accurate results and fair outcomes, we may need to consider the distribution of accuracy loss as a fairness concern across the users of the system (Patro et al. 2020).

The picture changes when we consider *consumer-side fairness*. Here we are interested in fairness considerations across the end users themselves, and this requires a community orientation in how the recommendation task is understood. We can draw on employment as an example yet again, now in terms of recommending jobs to users.

The tension between personalization and other goals becomes complex when we consider that users' behaviors themselves, the raw data over which personalization operates, may themselves be subject to measurement inconsistency. For example, female users are known to be less likely to click on ads that contain masculinized language about job performance (e.g., "rock star programmer"), but this says nothing about their underlying capabilities for such jobs (Hentschel et al. 2014). Even more fundamentally, there may be differences among users in their experience of the data gathering required by personalized systems; individuals experiencing disempowerment may identify the surveillance capacities that enable personalization as yet another avenue of unwanted external control (V. I. Burke and R. D. Burke 2019).

Even if we postulate that profiles can be collected in a fair and acceptable manner, it still may be the case that a system performs better for some classes of users than others. Improving fairness for disadvantaged groups may involve lower performance for others, especially in a rivalrous context like employment, where, as noted above, recommending something to everyone is not desirable. For example, a recommender system might optimize for fairness in such a way that a particularly desirable job is shown more often to members of a disadvantaged group and less often to others. How should a user think about bearing some of the burden in terms of lower utility of providing fairer recommendations to other users in the system? Accepting such behavior in a system requires adopting a pro-social orientation toward the community of fellow platform users, something that may not be easy to cultivate in a multistakeholder recommendation context.

Finally, we should note that the perspectives of recommendation consumers and item providers do not exhaust the set of stakeholders impacted by a recommender system. Authors such as Pariser (2011) and Sunstein (2018) have noted the way in which algorithmic curation of news and information has potential far-reaching impacts on society and politics. Incorporating this wide set of stakeholders draws the focus of a recommender system even further from personalization as a defining characteristic.

3 Conclusion

This discussion shows that the need to incorporate fairness into personalization systems requires more than just technical intervention. The way users think about these systems will need to change radically, and the onus lies on technologists to provide new terminology and new narratives that support this change. A key step may be to acknowledge the multistakeholder nature of existing commercial applications in which recommendation and personalization are embedded and to challenge the simplistic user-centered narratives promulgated by platform operators.

Acknowledgments This work was supported by the National Science Foundation under Grant No. 1911025.

References

Abdollahpouri, Himan et al. (2020). "Multistakeholder recommendation: Survey and research directions". In: *User Modeling and User-Adapted Interaction* 30, pp. 127–158.

Barocas, Solon and Andrew D Selbst (2016). "Big Data's Disparate Impact". In: *California law review* 104.3, p. 671-732.

Baumer, Eric PS and Jed R Brubaker (2017). "Post-userism". In: *Proceedings of the 2017 CHI Conference on Human Factors in Computing Systems*, pp. 6291–6303.

Burke, Robin (July 2017). "Multisided Fairness for Recommendation". In: Workshop on Fairness, Accountability and Transparency in Machine Learning. arXiv: 1707.00093 [cs.CY].

Burke, Victoria I and Robin D Burke (2019). "Powerlessness and Personalization". In: *International Journal of Applied Philosophy* 33.2, pp. 319–343.

DeVito, Michael A, Darren Gergle, and Jeremy Birnholtz (2017). ""Algorithms ruin everything" #RIPTwitter, Folk Theories, and Resistance to Algorithmic Change in Social Media". In: *Proceedings of the 2017 CHI conference on human factors in computing systems*, pp. 3163–3174.

Dwork, Cynthia et al. (2012). "Fairness Through Awareness". In: *Proceedings of the 3rd Innovations in Theoretical Computer Science Conference*. ITCS '12. New York, NY, USA: ACM, pp. 214–226.

Freeman, R Edward (2010). *Strategic management: A stakeholder approach*. Cambridge University Press.

Hentschel, Tanja et al. (Jan. 2014). "Wording of Advertisements Influences Women's Intention to Apply for Career Opportunities". In: *Academy of Management Proceedings* 2014.1, p. 15994.

Kendi, Ibram X (2019). *How to be an antiracist*. One World.

Liu, Weiwen et al. (2019). "Personalized Fairness-aware Re-ranking for Microlending". In: *Proceedings of the 13th ACM Conference on Recommender Systems*. RecSys '19. ACM.

Mehrotra, Rishabh et al. (2018). "Towards a fair marketplace: Counterfactual evaluation of the trade-off between relevance, fairness & satisfaction in recommendation systems". In: *Proceedings of the 27th ACM International Conference on Information and Knowledge Management*, pp. 2243–2251.

Mitchell, S., Potash, E., Barocas, S., D'Amour, A. and Lum, K., 2021. Algorithmic Fairness: Choices, Assumptions, and Definitions. *Annual Review of Statistics and Its Application*, 8, pp. 141-163.

O'Neil, Cathy (2016). *Weapons of math destruction: How big data increases inequality and threatens democracy*. Crown.

Pariser, E. (2011). *The filter bubble: How the new personalized web is changing what we read and how we think*. Penguin.

Patro, Gourab K et al. (2020). "Fairrec: Two-sided fairness for personalized recommendations in two-sided platforms". In: *Proceedings of The Web Conference* 2020, pp. 1194–1204.

Sonboli, Nasim et al. (July 2020). "Opportunistic Multi-aspect Fairness through Personalized Re-ranking". In: *Proceedings of the 28th ACM Conference on User Modeling, Adaptation and Personalization*. UMAP '20. Genoa, Italy: Association for Computing Machinery, pp. 239–247.

Sunstein, C.R. (2018). *#Republic: Divided democracy in the age of social media*. Princeton University Press.

Part VI
Platform Power

The Curation Chokepoint

James Larus

Abstract A key rationale for Apple and Google's app stores was that they curate apps to ensure that they do not contain malware. Curation has gone beyond this goal and now unduly constrains the apps that you can use on your smartphone. It needs to stop. App quality should be ensured with other techniques and by a wider range of organizations than just Apple and Google.

Imagine if you can a dystopia in which your landlord decides the food you can bring home to your apartment, either to cook or eat. Or that a manufacturer decides if a movie will play on your TV. Further imagine that your landlord and TV manufacturer demanded 30% of your grocery budget and Netflix subscription price for this "service."

This scenario seems absurd. However, it is the current situation on your smartphone. Apple and Google are a duopoly that wrote the low-level software (the operating systems iOS and Android) that control virtually all smartphones throughout the world. Apple and Google decide which apps can be installed on your smartphone, and they charge for this service.

Both companies operate an online "app store" that is the chokepoint in the distribution of apps. Apple iPhones can *only* install apps distributed by Apple's App Store.[1] Google permits alternative app stores but heavily promotes and favors its Play Store (e.g., it is pre-installed and using another app store may require changing a setting). Both companies' app stores heavily curate the apps they accept, and both charge a 30% commission on purchase of apps and on subsequent purchases performed within the apps themselves.

 Neither company's app store makes any pretense of being a neutral marketplace. Listing an app in either store requires agreeing to a contract with precise terms

---

[1] Apple makes an exception to allow companies to write and distribute software on smartphones that they own.

J. Larus (✉)
EPFL, Lausanne, Switzerland
e-mail: James.larus@epfl.ch

H. Werthner et al. (eds.), *Perspectives on Digital Humanism*,
https://doi.org/10.1007/978-3-030-86144-5_21

specifying the allowed types of apps, permissible software development techniques, and details of how an app operates. Companies developing apps have long chaffed at the heavy-handed and anti-competitive operation of these stores, but they have done so quietly for fear of retaliation. This concern is real and based on both companies' heavy-handed ways of resolving disputes, which start by using the vast disparity in negotiating strength to remove a contested app from the app store, thereby cutting off a developer's revenue. Recently, emboldened by governmental investigations in the USA and Europe, some companies have gone public with their disputes and complaints:

- Spotify, the Swedish online music streaming service, filed an anti-trust complaint with the EU, accusing Apple of unfair competition because it charges Spotify 30% of its subscription revenues. Apple distributes its competing Apple Music app pre-installed on iPhones without a fee (Satariano and Nicas 2019).
- The Swiss privacy software firm ProtonMail complained about the 30% fee and also that Apple insisted it remove the statement that its VPN app "unblock [s] censored websites" (Yen 2020). They further complained that Apple threatened to remove their email client app from the App Store unless they added an in-app purchase of email accounts, so that Apple could charge a commission. After agreeing to Apple's demand, ProtonMail raised the price of their email by 30%.
- Apple rejected an app that monitored Tesla cars' information because it used an unofficial library to extract data from a car. Apple claimed that the app developer needed approval from Tesla to use this library in its app (Espósito 2020).
- The developer of the popular Fortnite game, Epic Games, filed a lawsuit against Apple for anti-trust violations for requiring the use of Apple's payment system for in-game purchases, with its 30% commission (Nicas 2021). After the suit, Apple removed Fortnight from its App Store. A similar complaint against Google led to Fortnite's removal from Google's Play Store and a similar lawsuit.
- Facebook added a COVID-related service that allowed Facebook users to purchase online classes, with all revenue going to the business. Google allowed this, but Apple charged a 30% commission and prohibited the app when Facebook added a note informing the consumer that "Apple takes 30% of this purchase" (Lee 2020).
- Apple rejected gaming apps from Facebook and Microsoft because they were arcades that allowed subscribers to download and play games not installed through the App Store.

While considerable public attention has focused on the 30% commission and anti-competitive practices of the app stores, this chapter focuses on a different, fundamentally harmful consequence of Apple and Google's app stores. A key rationale for these stores is that both companies curate apps to ensure that they do not contain malware (software that subverts a computer or steals information). This vetting process is valuable and practical, as malware is far rarer on smartphones than on computers.[2]

Detecting malware is technically challenging. The fields of computer security and program analysis have developed numerous techniques for identifying malevolent programs, but the most active and creative malicious adversaries are typically one step ahead. Apple's rules for submitting apps to the App Store have a clear intent of facilitating code inspection, for example, by limiting apps to using published APIs and libraries, not downloading libraries or code, and not using interpreted code. The inspection process details are confidential, but Apple's documentation suggests it consists of automatic inspection of an app's code and manual execution to explore and approve its behavior.

While Apple's efforts may have raised most apps' quality level, it has not deterred sophisticated hacking groups such as NSO, whose spyware has been used by governments to track activists (Kirchgaessner and Safi 2020; Wolff 2019). NSO's Pegasus spyware used flaws in Apple's iMessage app (pre-installed) and Facebook's WhatsApp (App Store) to install spyware used by Middle Eastern countries to track political opponents and reporters. It is not surprising that curation is imperfect; one of the first and most profound theoretical results in computer science, Turing's Halting Problem, established that it is impossible to prove most non-trivial properties of computer programs. Program defect detection and security analysis rely on approximate analysis, which inherently suffers from false positives and missed errors.

The legal discovery process in Epic's lawsuit against Apple documented that the review process was cursory. In 2016, reviewers spent 13 min per new app (6 min per updated app) and were expected to review 50–100 apps per day (Epic Games 2021). The *Financial Times* quoted Eric Friedman, head of [Apple's] Fraud Engineering Algorithms and Risk (Fear) unit, that the process was "more like the pretty lady who greets you ... at the Hawaiian airport than the drug-sniffing dog" and assessed the App Store's defenses against malware as "bringing a plastic butter knife to a gunfight" (Chung 2021).

It turns out that, once malware is installed on an iPhone, Apple's strong isolation and restrictive rules shield it by, paradoxically, preventing the creation and distribution of effective anti-virus protection apps for iOS (O'Neill 2021).

Moreover, malicious adversaries can take advantage of Apple's ahead-of-time approval process, which examines an app before distribution and does not monitor its subsequent behavior on smartphones, to exhibit one face to Apple and another, less benign one to users (a violation of Apple's license agreement, but a malware vendor need not abide a license).

---

[2] Apple, in its Proposed Findings of Fact and Conclusions of Law, in Epic's lawsuit asserts: "the iPhone platform accounted for just 0.85% of malware infections. DX-3141 at 15. By contrast, Android accounted for 47.15% and Windows/PC accounted for 35.82%" (Apple 2021).

Apple's approach of claiming exclusive control over security and depending on App Store curation and isolation mechanisms on iPhones runs against the grain of centuries of security experience demonstrating the need for defense-in-depth.

More importantly, why should Apple determine which software is innocuous for all consumers who purchase an iPhone? Perhaps my risk tolerance is higher than average, and I want to try apps from developers who push the boundaries of what is possible on a phone and need to use libraries and techniques that Apple cannot inspect and so prohibit. Or perhaps my aesthetic sensibilities differ from the premise of the App Store, "The guiding principle of the App Store is simple—we want to provide a safe experience for users to get apps .... We have lots of kids downloading lots of apps...." Or perhaps I am a developer who runs afoul of Apple's self-admittedly vague and arbitrary standards:

> We strongly support all points of view being represented on the App Store, as long as the apps are respectful to users with differing opinions and the quality of the app experience is great. We will reject apps for any content or behavior that we believe is over the line. What line, you ask? Well, as a Supreme Court Justice once said, "I'll know it when I see it." And we think that you will also know it when you cross it. (Developer n.d.)

In the end, the fundamental question is whether the manufacturer of *my* phone should have the right to deny *me* the ability to install an app on *my* phone, and, conversely, the right of a software developer to produce and distribute an app, even if Apple finds the app inappropriate or offensive. I think that most reasonable people would say no.

Apple provides a valuable service by offering a curated collection of apps, much as Disney provides a service by providing entertainment appropriate for an entire family. However, no one, including Disney Corporation, would contend that their products encompass the full spectrum of entertainment or satisfy all tastes. Nor does it need to, as there are many other ways in which movies and television are produced and distributed.

The valuable aspect that Apple offers (malware protection) is achievable in other ways. Google's Android operating system allows alternative app stores. It is easy to envision an ecosystem of app stores, which offer software along with guarantees of its provenance (e.g., it is produced by a small firm that we know), application of an App Store-like inspection process, or even stronger techniques of program analysis (e.g., we inspected the source code of the application and built it ourselves). In many other domains, quasi-public organizations (e.g., UL in the USA or CE certification in Europe[3]) attest that a product meets publicly approved standards.

Moreover, Apple should not rely on a single mechanism (ahead-of-time App Store inspection) to provide security, as demonstrated by NSO malware. The "sandbox" mechanism on the iPhone, which isolates an app and controls the smartphone features it can access, needs further strengthening and restructuring to

---

[3] Underwriters Labs (UL) is a company that provides safety testing and certification of products, particularly for the USA. CE certification indicates that a product sold in the European Union conforms to EU health, safety, and environmental protection standards.

allow a phone's user to control over its operation. A sandbox can prevent an app from accessing the user's private information, such as their address book, using privacy-revealing mechanisms such as GPS, or communicating outside of the phone with a radio or WiFi. It is a powerful mechanism for controlling what runs on smartphones, too powerful to be left entirely in Apple and Google's hands.

A phone's owner should use these mechanisms to impose restrictions that conform to their desired risk level and expected behavior. A 15-year-old teenager may want to try edgy new apps and not be particularly concerned about personal privacy. A 45-year-old CEO is likely to be genuinely concerned about their work phone's security but more relaxed about their personal phone. One size does not fit all.

Apple and Google provide a valuable service by offering carefully inspected apps. This initial motivation made the app stores and smartphones successful and provided both companies with powerful commercial leverage to control and financially exploit the companies that write software for their phones. As governments increasingly examine these practices, it is essential not to lose sight of these app stores' stated motivation. Curation to exclude malware can be done in many ways and by many parties, and curation by content is only justified if alternative distribution mechanisms exist and are equally accessible.

References

Apple. 2021. "Apple Inc.'s Proposed Findings of Fact and Conclusions of Law." Case 4:20-cv-05640-YGR. https://www.scribd.com/document/502037049/21-04-07-Apple-Proposed-Findings-of-Fact-and-Conclusions-of-Law#from_embed.

Chung, Jean. 2021. "Apple Engineer Likened App Store Security to 'Butter Knife in Gunfight.'" *Financial Times*, April 9, 2021.

Developer, Apple. n.d. "App Store Review Guidelines." Accessed February 19, 2021. https://developer.apple.com/app-store/review/guidelines/.

Epic Games. 2021. "Findings of Fact and Conclusions of Law Proposed by Epic Games, Inc." https://www.scribd.com/document/502036985/21-04-08-Epic-Games-Proposed-Findings-of-Fact-and-Conclusions-of-Law#from_embed.

Espósito, Filipe. 2020. "Apple Rejects 3rd-Party Tesla App Update as It Strictly Enforces Written Consent for Third-Party API Use." 9To5Mac. August 27, 2020. https://9to5mac.com/2020/08/27/apple-rejects-watch-for-tesla-app-as-it-starts-requiring-written-consent-for-third-party-api-use/.

Kirchgaessner, Stephanie, and Michael Safi. 2020. "Dozens of Al Jazeera Journalists Allegedly Hacked Using Israeli Firm's Spyware." *The Guardian*, December 20, 2020, sec. Media. http://www.theguardian.com/media/2020/dec/20/citizen-lab-nso-dozens-of-aljazeera-journalists-allegedly-hacked-using-israeli-firm-spyware.

Lee, T. B. (2020). Apple backs down on taking 30% cut of paid online events on Facebook. https://arstechnica.com/gadgets/2020/09/apple-backs-down-on-taking-30-cut-of-paid-online-events-on-facebook/

Nicas, J. (2021). Apple Fortnite Trial Ends with Pointed Questions and a Toast to Popeyes, https://thecollegesave.com/2021/05/25/apples-fortnite-antitrust-trial-ends-with-pointed-questions-2/.

O'Neill, Patrick Howell. 2021. "How Apple's Locked down Security Gives Extra Protection to the Best Hackers." *MIT Technology Review*, March 31, 2021. https://www.technologyreview.com/2021/03/01/1020089/apple-walled-garden-hackers-protected/.

Satariano, A and Nicas, J. (2019). Spotify Accuses Apple of Anticompetitive Practice in Europe, The New York Times.

Wolff, Josephine. 2019. "Whatever You Think of Facebook, the NSO Group Is Worse." *The New York Times*, November 6, 2019. https://www.nytimes.com/2019/11/06/opinion/whatsapp-nso-group-spy.html.

Yen, Andy. 2020. "The App Store Is a Monopoly: Here's Why the EU Is Correct to Investigate Apple." *ProtonMail Blog* (blog). June 22, 2020. https://protonmail.com/blog/apple-app-store-antitrust/.

Business Model Innovation and the Rise of Technology Giants

Geoffrey G. Parker

Abstract Technology giants owe much of their success to fundamental improvements in the science and technology of information and communications technology. However, complementary advancements were also necessary, and, much as firms had to learn to incorporate electricity in the last nineteenth and early twentieth centuries, we posit that the giant platforms have learned to harness the contributions of external actors in order to grow more rapidly than would otherwise have been possible. Thus, the drivers of the dramatic rise of the tech giant platform firm can be viewed as a business model innovation as well as a technical innovation. As orchestration business models become better understood, we expect that firms in non-platform sectors are increasingly likely to adapt practices that also allow them to participate in and benefit from external value creation. At the same time, we expect regulatory scrutiny to increase as the power and reach of technology giants increases and their influence is felt across the economy.

1 The Rise of Technology Giants

The rise of giant technology firms such as Amazon, Alibaba, Microsoft, and more has been widely observed and commented upon in the popular press as well as the academic literature. While such firms were once Silicon Valley and Shenzhen phenomena, in recent years they have exploded in their power and reach. The drivers of this rise stem first from dramatic improvements in physics and the material science of computation technology, digital storage technology, and communications technology. These improvements have facilitated the growth of the information economy with implications for nearly every sector of society.

Beyond the raw improvements in information and communications technology (ICT) hardware and associated software improvements, there have been corresponding changes in the ways that organizations use their ICT investments.

G. G. Parker (✉)
Dartmouth College, Hanover, PA, USA
e-mail: geoffrey.g.parker@dartmouth.edu

© The Author(s) 2022
H. Werthner et al. (eds.), *Perspectives on Digital Humanism*,
https://doi.org/10.1007/978-3-030-86144-5_22

At an individual level, the willingness to connect with one another using tools such as email and social media took years to emerge but has now become part of the fabric of everyday life. At the firm level, it took many years for organizations to incorporate improved ICT into their processes. Arguably, the productivity boom of the 1990s was driven by firm-level learning as described by Brynjolfsson and Hitt (2000): "As computers become cheaper and more powerful, the business value of computers is limited less by computational capability and more by the ability of managers to invent new processes, procedures and organizational structures that leverage this capability." This process of learning to use ICT effectively was foreshadowed by the length of time that it took society to adapt to electric power. This was described by Harford (2017) as follows. "Factories could be cleaner and safer—and more efficient, because machines needed to run only when they were being used. But you couldn't get these results simply by ripping out the steam engine and replacing it with an electric motor. You needed to change everything: the architecture and the production process." A quote from Roy Amara suggested that this delayed adaptation is general: "We tend to overestimate the effect of a technology in the short run and underestimate the effect in the long run."[1]

2 The Inverted Firm and the Spread of Platform Business Models

Similar to the way that firms had to learn to incorporate electricity in the last nineteenth and early twentieth centuries and then 100 years later learn to integrate ICT into their business processes in the 1990s, we posit that the giant platforms have learned to orchestrate the efforts of actors outside the firm so that they were able to grow much more quickly than they otherwise would have. The dramatic rise of the tech giant platform firm is as much a business model innovation as it is a technical innovation. This can be seen in the way that firms such as Airbnb are able to compete with traditional hotel chains such as Marriott despite owning no hotels and directly employing very few people. Similarly, firms such as Lyft and Uber entered taxi and limousine markets despite employing few people and owning minimal physical assets. In particular, platforms are able to facilitate value creating interactions between external producers, content providers, developers, and consumers (Constantinides et al. 2018).

The ecosystem innovation required new managerial capabilities as well as new economic thinking (Helfat and Raubitschek 2018). The decision of whether to vertically integrate as firms or to instead use markets has been a long running topic of the "markets and hierarchies" literature (Williamson 1975). Parker et al. (2017) introduced the term "inverted firm" to describe the shift from value creation

---

[1] https://www.oxfordreference.com/view/10.1093/acref/9780191826719.001.0001/q-oro-ed4-00018679

inside the firm to outside. Importantly, platform firms are an intermediate organizational form that lies between pure market and pure hierarchy. The platform organizational form echoes the organizational forms that rely heavily on outsourcing. The automotive industry provides some of the best documented examples of the heavy use of outsourcing, especially in the descriptions of Toyota and its tightly integrated supply chain partners (Womack et al. 1990).

Platforms have taken this idea of working with external partners and made it possible for many more to participate at much lower cost. To do this, they employ open architectures with published standards to enable different types of users to interact to exchange physical and digital goods as well as services (Constantinides et al. 2018). Platforms also invest in organizational capabilities to provide governance rules and the capacity to enforce those rules. The goal is to maintain control over the platform while maintaining incentives for ecosystem partners to participate. In addition, investments in governance help to foster market safety so that users feel secure that they will not be taken advantage of when making transactions on the platform (Roth 2007).

Platforms are also able to facilitate and benefit from network effects that are otherwise too small to matter. The widely discussed concept of network effects describes the situation where systems become more valuable to users as the number of users increases (Shapiro and Varian, 1998). For example, the ratings that users provide on Netflix or YouTube create benefits for other uses that are too small for any one individual to try to reward or capture; transaction costs would overwhelm the benefits from such interactions. By using common systems, platforms are able to aggregate and analyze ratings information and make them available to all users in the form of direct rankings and better matching and filtering of content to consumers.

3 Demand for Regulation

Almost inevitably, the rapid growth of the largest technology firms has spurred calls for the regulation of those firms. Concerns range over issues such as antitrust and abuse of dominance, privacy, fair compensation for data provided, the dissemination of false information, and the regulation of speech. The European Union has led the world by implementing regulations, such as GDPR (General Data Protection Regulation) and PSD2 (Revised Payment Services Directive), and has recently proposed the DMA (Digital Markets Act). A recent panel report from the European Commission Joint Research Centre analyzes the different issues that the DMA is designed to address (Cabral et al. 2021). The panel's goal was to explain the economics behind the DMA and to comment on the proposals. Under the proposal, all online intermediaries offering services and content in the EU, whether they are established in the EU or not, have to comply with the new standards. Small companies will have obligations proportionate to their ability and size, but also remain accountable.

One critical impact that the Cabral report notes is creation of "black lists" of prohibited platform activities and "white lists" of allowable practices. The panel proposed a "grey list" of activities that platforms might be able to argue are likely to benefit consumers more than are likely to do harm. The Cabral panel also focused considerable attention on data sharing and the obligation for gatekeeper platforms to allow equal data access to all market entrants at the point of collection—known as "in situ." Such a mechanism avoids the need for consumers, or firms that are authorized to access consumer data, to download data. Instead, an in situ mechanism brings algorithms to the data where it is collected and stored. This should foster both security and better access for potential entrants.

4 Conclusions

The rise of technology firms and platform business models has created considerable benefits for consumer and business users. However, the rise also poses enormous challenges for regulators, incumbent firms, and individuals. Much of what economists and business scholars have studied derives from the first wave which was largely a story of business-to-consumer (B2C) platforms. There are now signs that the next wave of investment will be directed toward the growth of business-to-business (B2B) platforms. This raises important questions over the ways that B2C and B2B platforms will have similar characteristics and along what dimensions they are mostly likely to differ.

References

Brynjolfsson, E. and Hitt, L.M., 2000. Beyond computation: Information technology, organizational transformation and business performance. *Journal of Economic perspectives*, 14(4), pp.23-48.

Cabral, L., Haucap, J., Parker, G., Petropoulos, G., Valletti, T. and Van Alstyne, M., 2021. The EU Digital Markets Act, Publications Office of the European Union, Luxembourg, ISBN 978-92-76-29788-8, doi:https://doi.org/10.2760/139337, JRC122910.

Constantinides, Panos, Ola Henfridsson, and Geoffrey G. Parker. 2018. "Introduction—Platforms and Infrastructures in the Digital Age." *Information Systems Research* 29, no. 2: 381–400. https://doi.org/10.1287/isre.2018.0794.

Harford, T. "Why didn't electricity immediately change manufacturing." *BBC World Service* 50 (2017). Retrieved on 5-April-2021 from https://www.bbc.com/news/business-40673694.

Helfat, Constance E., and Ruth S. Raubitschek. 2018. "Dynamic and integrative capabilities for profiting from innovation in digital platform-based ecosystems." *Research Policy* 47.8: 1391-1399.

Parker, Geoffrey, Marshall Van Alstyne, and Xiaoyue Jiang. 2017. "Platform Ecosystems: How Developers Invert the Firm." *MIS Quarterly* 41, no. 1: 255–66. https://doi.org/10.25300/misq/2017/41.1.13.

Roth, Alvin E. 2007. "The art of designing markets." *Harvard Business Review* 85.10: 118.
Shapiro, Carl, and Hal R. Varian. 1998 *Information rules: A strategic guide to the network economy.* Harvard Business Press.
Williamson, O.E., 1975. Markets and hierarchies: analysis and antitrust implications: a study in the economics of internal organization. *University of Illinois at Urbana-Champaign's Academy for Entrepreneurial Leadership Historical Research Reference in Entrepreneurship.*
Womack, J. P., Jones, D. T., & Roos, D. 1990. *The Machine that Changed the World.* Simon and Schuster.

Scaling Up Broken Systems? Considerations from the Area of Music Streaming

Peter Knees

Abstract We discuss the effects and characteristics of disruptive business models driven by technology, exemplified by the developments in music distribution and consumption over the last 20 years. Starting from a historical perspective, we offer insights into the current situation in music streaming, where technology has not only changed the way we access music but also has important implications on the broader ecosystem, which includes the consumers, the authors, the record industry, and the platforms themselves. The discussion points to potential benefits, as well as to the risks involved in the currently deployed systems. We conclude that the increased profitability of the disruptive business models in the music domain and beyond is largely generated at the expense of the providers of the goods or services being brokered. Using the platforms as a consumer further subsidizes their value and might lead to mono- and oligopolies. While technology allows companies to effectively scale up business, the resulting systems more often amplify existing injustices than mitigate them.

1 Introduction

Technological developments have always created opportunities for business. This is especially the case for disruptive technologies. In the past decades, we have increasingly seen a fascination of venture capital with businesses promising disruption in their area and therefore maximizing investment return. Some popular examples of this paradigm include platforms for retail (Amazon, Alibaba), accommodation (Airbnb, Booking.com), individual transportation (Uber, Lyft), and media distribution (Netflix, Spotify) or, e.g., fintech services (Klarna, TransferWise). Characteristic to these business models across all domains is some form of incentive, such as improved user experience and/or reduced cost, for the various stakeholders involved, and increased efficiency by automating central elements of the traditional business

P. Knees (✉)
Vienna University of Technology, Vienna, Austria
e-mail: peter.knees@tuwien.ac.at

© The Author(s) 2022
H. Werthner et al. (eds.), *Perspectives on Digital Humanism*,
https://doi.org/10.1007/978-3-030-86144-5_23

approach, therefore allowing to scale up easily. These advantages undoubtedly are a key element for the success of the new business forms and overall the Internet-driven economy. Looking at the old-fashioned, obviously imperfect, and ultimately replaced business models reveals many flaws and inherent injustices. However, comparing them with their disruptive successors, it shows that the stakeholders consisting of a group of individual, non-organized participants (the providers of goods, as well as the consumers) tend to further lose power and influence in favor of the business owners providing the service. This raises the question: are these digital technologies merely permitting to scale up systems that were broken from the beginning? And if so: are disruptive digital technologies just amplifying injustices of the prior status quo?

In the following, we ask this question from the perspective of the area of music distribution. Music played an early and prominent role in the digital transformation of media delivery and consumption and has undergone significant changes over the last 25 years from a brick-and-mortar retail good to a ubiquitously available, de facto free service. We aim at giving a historical perspective describing the situation and the different stakeholders involved prior and past the disruption of the business and give an account on the beneficiaries and benefactors of this process, before drawing parallels to other areas of business disruption.

2   A Brief History of Music Business

Historically, the development of music production, distribution, and consumption is tightly intertwined with the technical possibilities of their respective time. Consider, for instance, the consequences of being able to transmit, physically record, and play back acoustic signals and the resulting means of broadcasting music and decoupling performance and listening. Over the second half of the twentieth century, this has led to the establishment of a record industry that would shift focus and power away from the composers of music and the publishing of sheet music toward the performers and marketing of events and trends. Revenue in this system was mostly generated by selling physical records on various types of media, most prominent on vinyl, music cassette, and compact disc (CD), with various stakeholders (among others, composers, performers, record labels, publishers, producers, distributors, and retailers) ideally being remunerated according to their negotiated share. Other, comparatively minor sources of revenue would stem from royalty payments based on airplay, performance, and licensing, as well as concerts. Radio was (and is) primarily a marketing tool to promote artists and records. To ensure payment to the various stakeholders, royalty collecting agencies established themselves to act on behalf of the rights owners, setting up rules for money distribution based on measurable, albeit ultimately arbitrary criteria, such as chart positions and other industry-conceived mechanisms. Essentially, this yielded a division of the area of music into "superstars" that could easily claim a (large, actually overproportioned) share and those who would have to fight for and often yield theirs.

Overall, this system favored the record companies over the individual artists and creators, as controlling the music production means would allow them to exert power over artistic decisions and aggregate various licensing rights while maintaining strong ties to the media and other distribution channels. In practice, the market was controlled by the major record labels and a number of gatekeepers, establishing an oligopoly that works for a few at the core of the industry at the expense of the many contributors. As a side effect of this model, however, record companies would also afford to finance the development of new artists, i.e., to invest in artists and repertoire (A&R). For independent artists, the diversity of media outlets and distribution channels would nonetheless allow them to find niches and address potential audiences.

3   The Disruption of the Business

The end of the twentieth century brought new technological developments, again impacting the music industry—and again, before all others. The adoption of the Internet as communication medium and the invention of audio data compression algorithms and formats, specifically MP3, led to active exchange of digital music files among users via decentralized peer-to-peer networks such as Napster, outside all established channels, infrastructures, and monitoring mechanisms built by the music industry. Needless to say, the resulting, almost instant availability of comprehensive music repositories to every peer of the network posed a threat to the core business model of the industry. Instead of embracing the new technological opportunities, the developers and users of the networks were criminalized and the downfall of popular musicians and music as an artform itself foretold as the music industry would not be able to keep up the monetization strategy that supported all stakeholders, related businesses, and intermediaries involved. As a reaction, the entertainment industry successfully lobbied for changes in copyright and intellectual property rights laws to further strengthen their position.

A few years into these developments, which indeed brought declining profits for the music ecosystem in comparison to the peak of CD sales in 2001 (IFPI 2021), new players and services explored the potential of the informatization of music reproduction and the decoupling from a physical media in alternative business models. These ranged from traditional e-commerce models of digital stores for music files to newer paradigms such as flat-rate subscriptions or freemium models typically with an ad-based entry level for unlimited music streaming. Together with the rise of smartphones that pushed an "always online" philosophy, new forms of access to music were enabled.

Generally, this framework of digital distribution via online services offers many advantages over the historic approach and potential for mitigations to its shortcomings for the various stakeholders. For music consumers, instant, cheap, and legal access to large amounts of music becomes feasible. For music creators, including those who contribute to the production process, it can instantly provide traceable delivery of their music to their audiences without having to deal with the various gatekeepers and distribution hubs. For the industry in between, logistics of physical retail is no longer necessary. For the music platform itself, being the central hub for delivering music to users allows them to track music listening behavior and build personalized services such as recommender systems and advanced interfaces for music discovery (Knees et al. 2020). In this scenario, however, record labels might lose their position of power and be bypassed easily. Their bargaining chip to retain relevance in this development is the control over the back catalogues, i.e., licensing access to records by well-known and popular artists whose availability is imperative for broad adoption. Ultimately, the disruption of the business introduced streaming platforms as additional central market players, next to record companies, whereas brick-and-mortar retail has largely vanished.

4   The Status Quo

While sales of physical media have been declining since 2001, the rising revenue from streaming has almost compensated for the losses and delivered an overall turnaround in 2015, today constituting the largest source of revenue (IFPI 2021). With more than 440 million paid subscription users worldwide (ibid.), streaming is the dominating modality of (traceable) music consumption. Current catalogues of the market-leading streaming platforms such as Spotify, SiriusXM Pandora, Apple Music, Amazon Music, or YouTube Music offer up to 70 million music tracks (Spotify 2021). To sum up, music streaming has managed to scale up the music business both in terms of content accessible and users reached.

But how is this situation still (or even more?) profitable with more players to support? The payments made by streaming companies for individual streams are extremely small—on average about USD 0.004 per stream (see Pastukhov (2019))—and often even negligible in sum per artist. Moreover, this amount is paid to the owners of the master recordings, which are typically the record labels, which then pass on only a small fraction to the artists. A reason for this practice is that revenue from streaming is often not explicitly negotiated in contracts, particularly in those signed before the technology existed. Hence, record companies benefit from licensing their back catalogues as it allows them to keep the largest share of the royalty payments to themselves. Also, recent efforts to present more justified and "fair" money distributions models, e.g., by distributing subscribers' fees according to their individual listening preferences, as implemented in Deezer's user-centric payment system (Deezer 2021), are not improving the situation as again the money is not paid directly to the artists, but the producers and publishers. Furthermore, despite the technical possibility to trace all plays of individual tracks, information about play counts across various platforms remains opaque for music creators, especially on platforms driven by user-generated content like YouTube on which content identification focuses again on the needs of big rights owners and neglects independent artists. These shortcomings themselves led again to the establishment of disruptive businesses like Kobalt or UnitedMasters that combine the roles of independent labels, rights management, and publishing companies, providing easier access to platforms and increased transparency of the process in a unified interface.

Record companies can now focus their attention and marketing strategies to a few distribution channels instead of dealing with a variety of different outlets. Artists and repertoire (A&R) is not part of their core business anymore, as new trends and talents are observed on social media platforms and co-opted based on hypes. As streaming is not generating much revenue, alternative sources of income like live performances, merchandise, sponsoring, and licensing of tracks need to provide a stable financial foundation for artists. At the same time, record labels increasingly demand a share of this profit in exchange for managing and promoting artists.

While record labels have once more found their position at the center of the business at the expense of artists and music creators, the question of profitability of the streaming platforms remains. Based on the annual reports of the various platforms, it appears they are not (yet?) operating profitable businesses. Their value seems to lie elsewhere, namely, in the potential for commercially exploiting their user basis, not only by charging for the service, but in fact by providing a data basis for marketing analyses sold to brands and advertising (Zuboff 2018). This might explain why, for instance, Spotify is partly owned by an intricate structure of the biggest record companies and rights holders, consisting of Universal Music Group, Warner Music Group, Sony Music, and, most recent, Tencent (Ingham 2020; see also Eriksson et al. 2019).

5   ... And Beyond?

While in the music business the old industry reinvented their role and managed to ultimately shape the new music industry based on its assets and absorb the disruptive element, in other domains, disruption has led to a replacement of the intermediaries and gatekeepers. In the movie domain, Netflix, first as a recommender-driven mail-in service, then as a movie streaming pioneer, completely eliminated the established video and DVD rental business and opened up a market that is now a battleground for several competitors following the same model. In the transportation area, Uber entered a long-time undisputed market and took large shares by storm. In the tourism domain, platforms like Booking and Expedia pushed the traditional business model of travel agencies close to the edge of the cliff (while the COVID-19 crisis seemingly finally pushed it over).

While Web-based services are at the center of all these examples and provide a new dimension of user experience in all these domains, it is never the user side that makes the business profitable. In music, the changed industry landscape is carried by besting the composers and creators of music due to old contracts or reduced benefits. In transportation, the pressure is put on drivers to make up for reduced wages by increasing customer throughput. In the tourism domain, hotels are put under further pressure to reduce their own profit margins and undersell their capacities to get visibility on platforms. In the movie domain, on the other hand, the competition of streaming services and their newly founded studios seems to increase the opportunities within the movie and TV industry rather than forcing the producers and workers to produce more for less compensation. In this context, it should be noted that the movie industry, especially in the USA, is strongly unionized.

The user is not spared from possible negative consequences either. Resorting again to the music example, the status quo of having access to virtually all music all the time is a utopian situation for music aficionados, and the technical developments that have enabled this situation over the last decades are a shining example of the blessings of information technology. At the same time, users become strongly dependent on and locked into such platforms. On one hand, "having access" does not mean "to own," and "personal collections" consisting of playlists created on platforms might become incomplete or missing due to changed circumstances in licensing. On the other hand, such services become more indispensable the more they are used. This might lead to the more general observation that "the rise in market concentration is greater in industries that are more intensive users of digital technologies" (Qureshi 2019); i.e., the disruption of industries by means of digital technologies itself promotes the emergence of mono- or oligopolies. In the long run, this is disadvantageous for the user and most other stakeholders involved.

To conclude, technology drives and "sells" disruptive business models. The increased profit of these models is often and largely generated by exploiting the people who provide the actual value behind the product, especially if they are not organized to fight for their joint interests. The users further subsidize these businesses simply through usage. From the perspective of Digital Humanism, this situation is unsatisfactory. We should first use technology to overcome the many broken systems before helping scaling them up.

References

Deezer (2021). Deezer wants artists to be paid fairly. Online: https://www.deezer.com/ucps, retrieved: 05/02/2021.

Eriksson, M., Fleischer, R., Johansson, A., Snickars, P., & Vonderau, P. (2019). Spotify Teardown: Inside the Black Box of Streaming Music. Cambridge: MIT Press.

IFPI (2021). Global Music Report 2021. Online: https://gmr2021.ifpi.org/assets/GMR2021_State%20of%20the%20Industry.pdf, retrieved: 05/02/2021.

Ingham, T. (2020). Who Really Owns Spotify?. Rolling Stone. Online: https://www.rollingstone.com/pro/news/who-really-owns-spotify-955388/, retrieved: 05/02/2021.

Knees, P., Schedl, M., & Goto, M. (2020). Intelligent User Interfaces for Music Discovery. Transactions of the International Society for Music Information Retrieval, 3(1), pp.165–179. DOI: https://doi.org/10.5334/tismir.60

Pastukhov, D. (2019). What Music Streaming Services Pay Per Stream (And Why It Actually Doesn't Matter). Soundcharts. Online: https://soundcharts.com/blog/music-streaming-rates-payouts, retrieved: 05/02/2021.

Qureshi, Z. (2019). Inequality in the Digital Era. Work in the Age of Data. BBVA OpenMind Collection (12), BBVA.

Spotify (2021). Company Info. Online: https://newsroom.spotify.com/company-info/, retrieved: 05/02/2021.

Zuboff, S. (2018). The Age of Surveillance Capitalism: The Fight for the Future at the New Frontier of Power. London: Profile Books.

The Platform Economy After COVID-19: Regulation and the Precautionary Principle

Cristiano Codagnone

Abstract Online platforms are two-sided or multisided markets whose main function is matching different groups (of producers, consumers, users, advertisers, i.e., hosts and guest in Airbnb, audiences and advertised in Google, etc.) that might otherwise find it more difficult to interact and possibly transact. Some of the potential critical issues associated with the platform economy include the relationship between personhood (the quality and condition of being an individual person with protected sphere of privacy and intimacy) and personal data, on which the platform economy thrives by extracting behavioral surplus, scale to dominance and market power, and lockin for businesses. In this chapter, I first shortly review how the pandemic crisis has impacted the platform economy and what problems are being exacerbated. I then conclude and focus the core part of my analysis on the issue of regulation and particularly on the merits and limits of applying the precautionary principle when addressing the online platform economy.

1 Introduction

Online platforms are two-sided or multisided markets whose main function is matching different groups (of producers, consumers, users, advertisers, i.e., hosts and guest in Airbnb, audiences and advertised in Google, etc.) that might otherwise find it more difficult to interact and possibly transact. The economic theory explains how value generation and profit making rely on crossed network externalities: value on one side of the market depends on positive network externalities on the other side. In some cases, gratuity is the optimal price setting on one side of the market (e.g., Facebook). It also explains the specific nature of unpaid digital labor, that is, users providing "labor," in the case of Facebook in the form of content creation, without receiving a monetary remuneration. Platforms internalize these network externalities, by facilitating the matching between sides and reducing transaction costs. The

C. Codagnone (✉)

Dipartimento di Scienze Sociali e Politiche, Università degli studi di Milano, Milano, Italy

e-mail: Cristiano.codagnone@unimi.it

© The Author(s) 2022

H. Werthner et al. (eds.), *Perspectives on Digital Humanism*,

https://doi.org/10.1007/978-3-030-86144-5_24

central features of platforms are direct and/or indirect network effects. In platforms, more users beget more users, a dynamic which in turn triggers a self-reinforcing cycle of growth. Platforms represent a new structure for organizing economic and social activities and appear as a hybrid between a market and a hierarchy. They are match makers like traditional markets, but they are also company heavy in assets and often quoted in the stock exchange. Platforms, therefore, are not simply technological corporations but a form of "quasi-infrastructure." Indeed, in their coordination function, platforms are as much an institutional form as a means of innovation.

In my previous work on the topic (Bogliacino et al. 2020; Codagnone et al. 2019; Mandl and Codagnone 2021), I discussed some of the potential critical issues associated with the platform economy. First, there is the relationship between personhood (the quality and condition of being an individual person with protected sphere of privacy and intimacy) and personal data, on which the platform economy thrives by extracting behavioral surplus (Zuboff 2019). The extraction of personal behavioral data, including about things one may consider very personal and secrets, is a violation of personhood as defined above. In particular, "loss of control over personal information creates a variety of near-term and longer-term risks that are difficult for individuals to understand – and, importantly for antitrust purposes, therefore impossible for them to value" (Cohen 2019, p. 175). Access to such personal data enables companies to "hyper-nudge" consumers (Yeung 2017). Short-circuiting behavioral data and algorithmic learning, online platforms enact very powerful nudges guiding decisions and reducing consumers' autonomous choices. Second, there is the problem of competition and the potential monopolistic or oligopolistic outcomes of the platform economy, considering also the number of merger and acquisitions (M&A) completed by the most powerful of them (so-called GAFAM – Google, Amazon, Facebook, and Apple) between 2013 and 2019 (see, for instance, Lechardoy et al. 2021, pp. 46–47). Third, the platform economy, and especially online labor platform, is contributing to the fragmentation of work and to the rise of new precarious work forms (Mandl and Codagnone 2021).

In this chapter, using information mostly from the European Commission "Observatory on the Online Platform Economy" (https://platformobservatory.eu/ and the analytical paper by Lechardoy et al. 2021), I first shortly review how the pandemic crisis has impacted the platform economy and what problems are being exacerbated. I then conclude and focus the core part of my analysis on the issue of regulation and particularly on the merits and limits of applying the precautionary principle when addressing the online platform economy.

2   The Effects of the Pandemic

The COVID-19 pandemic has created a unique situation leading to an acceleration of digital transformation and digitization. The COVID-19 crisis has increased the use of online services and the breadth of users. A recent report by the Joint Research Centre (JRC, the research harm of the Commission) concludes that as a game changer,

COVID-19 is acting as a booster of both potential opportunities and of concerns/ risks for the digital transformation (Craglia et al. 2020). The opportunities are those of further innovation, economic growth, as well as increase in efficiency and effectiveness of public services. The risks include above all the deepening of the digital divide, the scale to dominance by incumbents, and the issue of personal data and surveillance. The acceleration in the pace of digitization has also exposed the opportunities and weaknesses of the online platform economy. For what concerns the latter, the main effects of the pandemic reported by the earlier cited Online Observatory (https://platformobservatory.eu/) and related analytical paper (Lechardoy et al. 2021) are four.

First, the Observatory surveys show that during the crisis business users have increasingly relied on online platforms despite the fact that the revenues generated through them decreased. There is little evidence of switching, which suggests the pandemic crisis has reinforced lockins.

Second, traffic and revenue share have increased for social media, search engines, and some national marketplaces, while they have decreased for platforms in the tourism and travel sectors. The top 5 platforms (Google, Apple, Facebook, Amazon, Microsoft) have increased their audience and recorded sizeable profits also in 2020. On the other hand, the pandemic accelerated the digital transition of health and education, where new platforms are emerging.

Third, the GAFAM incumbents have kept being active in M&A, purchasing companies to either strengthen their current products and services or expand into adjacent markets and consolidate their ecosystem. Since the pandemic has not weakened but rather reinforced the economic position of incumbent platforms, it is reasonable to expect that they will continue in their M&A activism in the near future. In view also of the fact that given the monetary threshold for M&A to attract regulators' attention, many M&A of small but innovative companies occur below the regulators' radar.

Fourth, many traditional business players (non-IT companies) are under dire condition of crisis due to the effect of the pandemic abilities to recover and to innovate. This may further reinforce the market position of platform payers that may appropriate new market shares.

Last but not least, online platform and related apps entered also directly into the business of tackling and containing the pandemic. Smart solutions enabled by online platform, has been argued, could improve detection and alerting (more effective and precise), as well as enable faster containment actions and leveraging positively existing infrastructures. So, dozens of online apps and platforms mushroomed for track and tracing. In April 2020, Apple and Google announced a partnership to enable the use of Bluetooth Low Energy technology for COVID-19 contact tracing, and in September 2020, Apple and Google announced that they would integrate their contact tracing technology into their next mobile operating systems claiming that health authorities would not need to build their own tracing apps but can simply configure the basic software framework to their country's needs. This development, however, has raised concerns on privacy and personal data and about the risk that, after the emergency would be over, our society could drift by inertia toward more systemic surveillance. Furthermore, it has been observed that the pandemic has helped incumbents such as Google to regain legitimacy and momentum (Cinnamon 2020).

3   Regulation and the Precautionary Principle

As reviewed, the effects of COVID-19 have increased some of the policy concerns surrounding the online platform economy. Concerns that before the pandemic outbreaks were high on the regulators agenda with a debate polarized basically among two positions.

The first is the libertarian and impossibility statement position. According to this view, any attempt to regulate online platform and more broadly the current digital transformation (including Artificial Intelligence, AI) would stifle innovation and produce undesirable side effects. In extreme fashion, this discourse can be summarized with the view that regulation is the mortal enemy of innovation (Cohen 2019, p. 178). Hence, for the sake of economic growth and innovation, matters should be deregulated, and/or their governance should be devolved to the private sector through various forms of self-regulation and de facto standardization. A corollary of this discourse is that attempts at regulation are touted as new forms of protectionism. A second discourse takes the form of an "impossibility statement." Regulation of current development is and will remain technically complex and beyond the reach of the cognitive tools and processes available to regulators. The impossibility statement implication is that in the age of algorithmic governance emerging as a new form of business strategy, regulators cannot keep up and should only hope and wait until algorithms improve and better self-regulate themselves.

The most sustained counterargument has been developed by law scholar Juliet Cohen who argues that the current digital transformation requires regulatory innovation not only on the "what" (new rubrics of activities needing regulation) but also on the "how," meaning entering the domain of algorithmic governance (2019, 812–185 and 200–201). Cohen and other scholars of the digital transformation are in favor of applying the precautionary principle to regulate, for instance, the way platforms gather personal data that generates a behavioral surplus. To a large extent, the difference between a precautionary approach to regulation and a cost-benefit one is how the object of analysis is positioned between the two poles of uncertainty and risk. Under high uncertainty and the possibility that no regulation would produce serious harms, then a precautionary approach would favor introducing regulation a priori to preempt such harms. The best analogy is with the introduction of lockdowns across entire population without any cost-benefit analysis based on the uncertain but serious danger of wider spread and more deaths from COVID-19.

So, given the complexity and uncertainty surrounding the development of platforms and related technologies, Cohen argues to move from a risk perspective backing a cost-benefit approach to introducing policy and regulation based on the precautionary principle. Competition regulation in the context of the digital transformation and with specific respect to online platforms requires a radical renewal of a regulatory regime that was developed for the industrial era and as such is no longer appropriate or useful. In sum, the view juxtaposed to the approach "leave it to the market" is that the current digital transformation requires a bottom-up rethinking of competition and the imposition of a public utility regulatory regimes on platforms. The possibility of imposing common carriage/public utility requirements on online platform can be justified in view of some of the considerations I placed in the introduction. Given their de facto role as "quasi-infrastructures" and "quasi-institutions," and also considering their prominence during the pandemic, this could be the good moment to adapt the industrial era notions of common carriage and/or public utility provision to the networked information age and particularly to online platforms. The argument goes that regulation of the digital world should incorporate public access and social justice considerations.

4   Conclusions

Without entering into the merits of the two opposing positions characterized above, I am going to conclude this chapter with a discussion of the application of the precautionary principle versus a case-by-case cost-benefit analysis as different approaches to regulation.

The application of the precautionary principle turns the complexity and uncertainty argument, often used by libertarian and/or tech lobbyists, on its head. According to Cohen (2019), for instance, platformization and algorithmic governance create such level of complexity and uncertainty that warrant moving away from a risk perspective backing a cost-benefit approach to policy and regulation to an uncertainty perspective backing a precautionary approach that would prescribe intervening with regulation. From this perspective, a precautionary approach should be adopted, and more stringent regulation enacted when uncertainties concern crucial and value-relevant issues. On the other hand, the precautionary approach has been criticized as "the law of fear" and considered inferior to cost-benefit analysis approach to policy issues on a case-by-case level (Sunstein, 2005). Although reasonable a priori, the precautionary principle is usually contested on two grounds: (a) if regulation is defended on the principle of the worst scenario, then a lack of regulation can be defended by the same argument when the consequences of strict regulations are potentially very negative; and (b) the precautionary principle claims that dangers should not be downplayed, but this exposes the risk of building a negative public discourse that would block innovators.

There is a point in Sunstein's critique of the precautionary principle, in that by reacting to uncertainty and complexity with across-the-board regulation may end up stifling true innovation without cutting the nails of the incumbents. There are many innovative platforms, and not all of them are or will become as GAFAM. The latter and the concerns they raise can only be dealt with new competition policy instruments and cases and with political will to do so. On the other hand, regulators should incentivize relevant actors to adopt governance standards and procedures that will support their efforts to operationalize trustworthy digital transformation and online platform economy. Furthermore, they should support the development of technologies, systems, and tools to help relevant actors identify and mitigate relevant risks. This means incentivizing organizations to adopt robust internal governance and equipping them with tools to identify and mitigate risk is considered more effective than a regulatory regime that mandates specific outcomes. New regulation should support ongoing efforts to build best practices, rather than risk cutting them short with inflexible rules that may not be able to adapt to a rapidly changing field of technology.

In conclusion, regulators should carefully weigh the pros and cons of policy responses adopting the precautionary principles and those that support a case-by-case cost-benefit analysis before introducing any new piece of legislation.

References

Bogliacino, F., Codagnone, C., Cirillo, V., & Guarascio, D. (2020). Quantity and quality of work in the platform economy. In K. Zimmermann (Ed.), Handbook of Labour, Human Resources and Population Economics. London: Springer Nature.

Cinnamon, J. (2020). Platform philanthropy, 'public value', and the COVID-19 pandemic moment. Dialogues in Human Geography, 10 (2), pp. 242-245.

Codagnone, C., Karatzogianni, A., & Matthews, J. (2019). Platform Economics: Rhetoric and Reality in the 'Sharing Economy'. Bingley, United Kingdom: Emerald Publishing Limited.

Cohen, J. (2019). Between Truth and Power: The Legal Constructions of Informational Capitalism. Oxford: Oxford University Press.

Craglia, M. et al. (2020). Artificial Intelligence and Digital Transformation: early lessons from the COVID-19 crisis. Luxembourg: Publications Office of the European Union.

Lechardoy, L., Sokolyanskaya, A. & Lupiáñez- Villanueva, F. (2021). Analytical paper on the structure of the online platform economy post COVID-19 outbreak. Brussels: Observatory on the Online Platform Economy, European Commission.

Mandl, I., & Codagnone, C. (2021). The Diversity of Platform Work— Variations in Employment and Working Conditions. In H. Schaffers, M. Vartiainen, & J. Bus (Eds.), Digital Innovation and the Future of Work (pp. 177-195). Gistrup, Denmark: River Publishers.

Sunstein, C. (2005). The Law of Fear. Cambridge: Cambridge University Press.

Yeung, K. (2017). 'Hypernudge': Big Data as a mode of regulation by design. Information, Communication & Society, 20(1), 118-136.

Zuboff, S. (2019). The Age of Surveillance Capitalism. London: Profile Books Ltd.

Part VII
Education and Skills of the Future

Educational Requirements for Positive Social Robotics

Johanna Seibt

Abstract Social robotics does not create tools but social 'others' that act in the physical and *symbolic* space of human social interactions. In order to guide the profound disruptive potential of this technology, social robotics must be repositioned—we must reconceive it as an emerging interdisciplinary area where expertise on social reality, as physical, practical, and symbolic space, is constitutively included. I present here the guiding principles for such a repositioning, "Integrative Social Robotics," and argue that the path to culturally sustainable (value-preserving) or positive (value-enhancing) applications of social robotics goes via a *redirection* of the humanities and social sciences. Rather than creating new educations by disemboweling, the humanities and social sciences, students need to acquire full disciplinary competence in these disciplines, as well as the new skill to direct these qualifications toward membership in multidisciplinary developer teams.

So-called social robots are artificial agents designed to move and act in the physical and symbolic space of human social interactions—as automated guides, receptionists, waiters, companions, tutors, domestic assistants, etc. According to current projections, already by 2025 there will be a US$100 billion market for service robots, and by 2050 we might have automated 50% of all work activities (McKinsey 2017). As economists gladly usher in the "automation age" (ibid.), it is crucial to be clear on a decisive difference between digitalization and unembodied AI on the one hand and embodied social AI on the other. The 'as if' of simulated sociality in embodied AIs (social robots) captivates us with disconcerting ease. For the first time in human cultural history, we produce, for economic reasons, *technological artifacts that are no longer tools* for us—we are building "social others".

More than a decade of human-robot interaction (HRI) research reveals how willingly humans engage with social robots, practically but also at the affective

J. Seibt (✉)
Aarhus University, Aarhus, Denmark
e-mail: filseibt@cas.au.dk

level, and these research results raise far-reaching theoretical and ethical questions. Should the goings-on between humans and robots really *count as* social actions? Will we come to prefer the new "friends" we *made* to human friendships we need to cultivate? If robots display emotions, which increases the fluidity of social interactions (Fischer 2019), will we be able to learn *not* to respond with moral emotions (sympathy)? Or should robots have rights? Will social robots de-skill us for interacting authentically with other people?

Decisions pertaining to the use of social robots are not only highly complex—as the example of sex robots may illustrate most strikingly—but also bound to have momentous socio-cultural repercussions. However, research-based policy making on social robots is currently bogged down in a "triple gridlock of description, evaluation, and regulation" (Seibt et al. 2020a, b) combining descriptive and prescriptive uncertainty. Currently, we do not know precisely how to describe human reactions to robots in non-metaphorical ways; the lack of precise and joint terminology hampers the comparability of empirical studies; and the resulting predictive uncertainties of our evaluations make it impossible to provide sufficiently clear and general regulatory recommendations.

Robotics engineers increasingly appreciate that their creations require decisional competences far beyond their scientific educations. Single publications (see, e.g., Nourbakhsh 2013; Torras 2018) and the recent IEEE "Global Initiative on Ethics of Automated and Intelligent Systems" document impressive efforts "to ensure every stakeholder involved in the design and development of autonomous and intelligent systems is educated, trained, and empowered to prioritize ethical considerations so that these technologies are advanced for the benefit of humanity" (IEEE n.d.).

While we should wholeheartedly endorse these efforts, the question arises whether new educational standards, requiring obligatory ethics modules in engineering educations, will be sufficient. More precisely, these efforts may not suffice as long as we retain the current model of the research, design, and development process (RD&D model) in social robotics.

According to the current RD&D model, roboticists, supported by *some* relevant expertise from other disciplines, create an *object* (robot) which is supposed to *function* across application contexts. What these objects *mean* in a specific application context hardly comes into view. Even if engineering students were to acquire greater sensitivity for the importance of ethical considerations—e.g., as a first step, the insight that "ethical considerations" go beyond research ethics (data handling, consent forms, etc.)—it is questionable that even a full year of study (using the European nomenclature: a 60 ECTS module) could adequately impart competences for responsible decision-making about the symbolic space of human social interactions. *The symbolic space of human interactions is arguably the most complex domain of reality we know of*—structured not only by physical and institutional conditions but also by individual and socio-cultural practices of "meaning making," with dynamic variations at very different time scales. Even a full master's study (4–5 years) in the social sciences or the humanities is barely enough to equip students with professional expertise (analytical methods and descriptive categories) necessary to understand small regions or certain aspects of human social reality.

In short, given that the analysis of ethical and socio-cultural implications of social robotics applications requires *professional expertise* in the social sciences or humanities, which short ethics modules cannot provide, our current RD&D model for the development of social robotics applications places responsibilities on the leading engineers that they cannot discharge.

The way forward is thus to modify the RD&D model for social robotics applications. In line with design strategies such as "value-sensitive design" (Friedman et al. 2002), "design for values" (Van den Hoven 2005), "mutual shaping" (Šabanović 2010), and "care-centered value-sensitive design" (Van Wynsberghe 2016), the approach of "Integrative Social Robotics" (ISR) (Seibt et al. 2020a, b) proposes a new developmental paradigm or RD&D model that is tailormade for our current situation. As a targeted response to the triple-gridlock and the socio-cultural risks of social robotics, ISR postulates an RD&D process that complies with the following five principles (for details, see ibid.):

> (P1) The *Process Principle*: The product of a RD& D process in social robotics are not objects (robots) but social interactions.

This principle makes explicit that social robotics generates (not instruments but) new sorts of interactions that (1) are best understood as involving forms of asymmetric sociality and, even more importantly, (2) belong into a complex network of human social interactions. This shift in our understanding of the research focus of social robotics immediately motivates the following principle:

> (P2) The *Quality Principle*: The RD&D process must involve, from the very beginning and throughout the entire process, expertise of all disciplines that are directly relevant for the description and evaluation of the social interaction(s) involved in the envisaged application.

The Quality Principle demands the constitutive involvement (a) of expertise pertaining to the functional success of the envisaged application—e.g., health science, gerontopsychology, education science, or nursing—but also (b) of expertise from researchers from the social sciences and humanities, in order to ensure adequate analyses of the socio-cultural interaction context of the application. Robotics engineers currently develop applications often guided by their own ethical imagination and social competence alone, which is, in our current situation, a highly problematic underestimation of the tasks and risks involved. One central condition for a deeper understanding of the socio-cultural interaction context is formulated in the following principle:

> (P3) The *Principle of Ontological Complexity*: Any social interaction I is a composite of (at least) three components <I1, I2, I3>, which are the agent-relative realizations of interaction conceptions as viewed from the perspective of (at least) two interacting agents and an external observer. The interaction conceptions of each of the directly interacting agents (i.e., I1 and I2) consist in turn of three perspectival descriptions (from a first-, second-, and third-person view) of the agent's contribution, while the interaction conception of the external observer is only from a third-person perspective. The RD&D process in social robotics must envisage, discuss, and develop social interactions on the basis of a perspectival account of social interactions.

A differentiated understanding of social interactions along these lines allows us to analyze in detail how people experience their interactions with a robot; in-depth investigations into these partly pre-conscious processes of "sociomorphing" and their associated conscious phenomenology (Seibt et al. 2020b) are crucial elements in the continuous evaluation of the robot's (physical, kinematic, and functional) design, as demanded by the following principle:

(P4) The *Context Principle*: The identity of any social interaction is relative to its (spatial, temporal, institutional, etc.) context. The RD&D process must thus be conducted with continuous and comprehensive short-term regulatory feedback loops (participatory design) so that the new social interaction is integrated with all relevant contextual factors. Importantly, participatory feedback continues for quite some time after the "placement of the robot," until a new normality has formed.

Since according to ISR Principle 2, the Quality Principle, researchers from the humanities and, in particular, ethicists are included in the RD&D process, the Context Principle ensures that both individual preferences of the stakeholders are taken into account as well as the interests of society at large. The Context Principle acknowledges the complexity of social reality, but also expresses a commitment to a combined empirical (bottom-up) and normative (top-down) determination of "what matters" in the given application context. This is reinforced by the following principle:

(P5) The *Values-First Principle*: Target applications of social robotics must comply with a specification of the *Non-Replacement Maxim*: *social robots may only do what humans should but cannot do*. (More precisely: robots may only afford social interactions that humans should do, relative to value V, but cannot do, relative to constraint C.) The contextual specification of the Non-Replacement Maxim is established by joint deliberation of all stakeholders. Axiological analyses and evaluations are repeated throughout all stages of the RD&D process.

The formulation of the *Values-First Principle*, as well as of the Non-Replacement Maxim, reflect a commitment to an account of values in line with classical pragmatism, values are realized in interactions and experiences, in a strictly context-dependent fashion. Accordingly, the ISR approach begins with an extended intake phase with careful field research on the value landscape of the interaction context before inserting the new interaction and is accompanied by an ongoing value dialogue of all stakeholders (including the ethicists among the researchers/developers) that reaches into the phase of the "new normal" many months after the placement of the robot.

The Value-First Principle of ISR can be applied with two different levels of ambition. First, applications can be selected that comply with the Non-Replacement Maxim relative to a given context C and raise a value V1 that within that context C ranks higher than a value V2 which is negatively affected by the application. For example, a companion robot for elderly residents of a nursing home which assists in establishing telecommunication with the resident's family will raise the autonomy of the resident; given that the nursing home cannot offer continuous human assistance *and* an in-person visitation plan with the family is in place, the resident may rank the increase in her autonomy higher than the loss of some human contact with staff.

Such an application, which warrants an increase of a value relative to the axiological ranking within the context, can be considered as *culturally sustainable*.

Second, working with a more restrictive selection filter, developer teams working with ISR might choose only to pursue applications that (1) comply with the Non-Replacement Maxim relative to a *class* of contexts and (2) raise a value V1 without affecting the axiological ranking in this context class. For example, delivery of conflict mediation via a genderless humanoid telecommunication robot can apparently increase the likelihood that conflict parties find constructive resolutions for gender-charged conflicts (Druckman et al. 2020); this seems to be due to the fact that the genderless robot does not provide stimuli for gender-related perceptual bias, a feature that human mediators should but (typically) cannot exhibit (Skewes et al. 2019). Such an application, where a human-robot interaction is *intrinsically valuable*, can be considered an instance of *positive social robotics* (in extension and further specification of the notion of "positive computing" (Calvo and Peters 2014)).

I have described the five principles of the ISR approach in some detail here in order to motivate the following three claims about what culturally sustainable and positive social robotics would imply for future higher education:

> *Claim 1: Reposition social robotics:* Given the currently incalculable socio-cultural risks of widespread use of social robots, we need new RD&D models (such as ISR) where (a) the expertise of the humanities and social sciences is centrally and constitutively included from beginning to end, and (b) applications are developed in a value-driven fashion.
>
> *Claim 2: No new educations, refurbish and redirect the humanities:* We need to conceive of social robotics as an emerging interdisciplinary area of research—but we are not yet in the position to introduce new bachelor's or master's educations in this area. Culturally sustainable or even positive social robotics currently still needs the combination of *professional* expertise in all relevant disciplines, i.e., expertise that is acquired in the course of full professional educations in these disciplines. However, if humanities educations in the future aim to qualify students for work in social robotics developer teams, the humanities also need to change their self-image from purely reflective to (also) pro-actively engaged areas of knowledge and adjust curricula accordingly.
>
> *Claim 3: Create interfaces*: In order to equip students with suitable skills for work in interdisciplinary developer teams for culturally sustainable or positive social robotics, we need to establish *interface modules*. The core disciplines of the emerging interdisciplinary area of social robotics, such as robotics, anthropology, philosophy, (geronto)psychology, design, health, nursing, and education, should include modules of the course work of one semester (typically three courses) where students acquire skills in interdisciplinary communication and collaboration, as well as some basic introductions to terminology and methods of other core disciplines, with focus on ethics.

Claim 1, postulating the constitutive inclusion of the humanities and social sciences, is a direct consequence of the second principle of ISR, the "Quality Principle": it is scientifically irresponsible to build high-risk applications without involving those disciplines that can professionally assess the risks of value corruptions (the corruption of public discourse and fact-finding practices by social media algorithms illustrates the consequences of irresponsible technology development).

Claim 3 follows from the fifth principle of ISR, the "Values-First Principle," which offers a gradual exit, piece-meal, from the current triple-gridlock of research-based regulation. If developer teams concentrate on positive, value-enhancing applications, we can gradually learn more about human encounters with social robots while reducing the potential risks—building the applications "we could want anyway." However, as my commentaries on the Values-First Principle may convey, value discourse and the analysis of values require a *certain mindset, a tolerance for ambiguity, complexity, and deep contextuality*, that cannot be learned by top-down rule application or as "how-to" lateral transfer. While an education in the humanities cultivates the development of such special mindsets, this is not so in other disciplines. A preparatory course in applied ethics with well-chosen examples (see, for instance, course material in connection with Torras 2018) can gradually train students from non-humanities disciplines to *understand* more about epistemic attitudes and standards of reasoning in complex normative domains, but it will likely not suffice to acquire them. Vice versa, as currently explored in a supplementary educational module on "Humanistic Technology Development" at Aarhus University, in order to create epistemological and communicational interfaces from the other side, humanities students should learn a bit of programming and design and build a rudimentary robot.

Finally, let us consider claim 2, the claim that we should better *not* push for new educations, at least not now. This claim might be surprising, in view of interdisciplines like bioinformatics or nanotechnology where the identification of a new research field led quickly to the introduction of new educations. On the other hand, as Climate Research and Systems Biology illustrate, there are interdisciplinary areas the complexity of which, relative to our current understanding, requires full disciplinary competences. If we begin to interfere with a domain as intricate as social reality, and care for more than money, we cannot afford cheap solutions with cross-disciplinary educations stitched together cut-and-paste. Given the complexity and contextuality of social reality, given the demand for value-driven applications based on quality research (see ISR Principles 2–5), social robotics needs developer teams with full expertise rather than mosaic knowledge in order to create futures worth living.

References

Calvo, R.A., Peters, D. (2014). Positive computing: technology for wellbeing and human potential. MIT Press.

Druckman, D., Adrian, L., Damholdt, M.F., Filzmoser, M., Koszegi, S.T., Seibt, J., Vestergaard, C. (2020). Who is Best at Mediating a Social Conflict? Comparing Robots, Screens and Humans. Group Decis. Negot. https://doi.org/10.1007/s10726-020-09716-9

Fischer, K. (2019). Why Collaborative Robots Must Be Social (and even Emotional) Actors. Techné Res. Philos. Technol. 23, 270–289.

Friedman, B., Kahn, P., Borning, A. (2002). Value sensitive design: Theory and methods. University of Washington technical report 02–12.

IEEE, n.d. IEEE SA – The IEEE Global Initiative on Ethics of Autonomous and Intelligent Systems [WWW Document]. URL https://standards.ieee.org/industry-connections/ec/autonomous-systems.html (accessed 10.28.20).

McKinsey Global Institute, A Future that Works, Automation, Employment and Productivity, 2017, https://www.mckinsey.com/mgi/overview/2017-in-review/automation-and-the-future-of-work/a-future-that-works-automation-employment-and-productivity

Nourbakhsh, I.R. (2013). Robot futures. MIT Press.

Šabanović, S. (2010). Robots in society, society in robots. Int. J. Soc. Robot. 2, 439–450.

Seibt, J., Damholdt, M.F., Vestergaard, C. (2020a). Integrative social robotics, value-driven design, and transdisciplinarity. Interact. Stud. 21, 111–144.

Seibt, Johanna, Vestergaard, C., Damholdt, M.F. (2020b). Sociomorphing, Not Anthropomorphizing: Towards a Typology of Experienced Sociality, in: Culturally Sustainable Social Robotics--Proceedings of Robophilosophy 2020, Frontiers of Artificial Intelligence and Its Applications. IOS Press, Amsterdam, pp. 51–67.

Skewes, J., Amodio, D.M., Seibt, J. (2019). Social robotics and the modulation of social perception and bias. Philos. Trans. R. Soc. B Biol. Sci. 374, 20180037.

Torras, C. (2018). The Vestigial Heart. MIT Press.

Van den Hoven, J. (2005). Design for values and values for design. Information age 4, 4–7.

Van Wynsberghe, A. (2016). Service robots, care ethics, and design. Ethics Inf Technol 18, 311–321.

Informatics as a Fundamental Discipline in General Education: The Danish Perspective

Michael E. Caspersen

Abstract Informatics in general, and the particular development of artificial intelligence, is changing human knowledge, perception, and reality, thus radically changing the course of human history. Informatics has made it possible to automate an extraordinary range of tasks by enabling machines to play an increasingly decisive role in drawing conclusions from data and then taking action. The growing transfer of judgment from human beings to machines denotes the revolutionary aspect of informatics.

For societies to maintain or regain democratic control and supremacy over digital technology, it is imperative to include informatics in general education with a dual perspective on possibilities and implications of computing for individuals and society. The Danish informatics curriculum for general education acknowledges the dual and bipartite nature of informatics by complementing a constructive approach to computing with a critical analytic approach to digital artifacts.

1 Digital Humanism and General Informatics Education

Information technology is a technology unlike any other humankind has invented. All other technologies stretch our *physical ability* by enabling us to move faster from one place to another, to generate (green) energy, to develop life-saving medicine, to refine food production, and so forth. Information technology is crucial for other modern technologies, but the essential unique quality of information technology is that it stretches our *cognitive ability*.

Already in 1967, Danish Turing Laureate Peter Naur wrote about the importance of including informatics in general education (Naur, 1967, pp. 14–15; Naur 1992, p. 176):

M. E. Caspersen (✉)
It-vest – Networking Universities, Department of Computer Science, Aarhus University, Aarhus, Denmark
e-mail: mec@it-vest.dk

© The Author(s) 2022
H. Werthner et al. (eds.), *Perspectives on Digital Humanism*,
https://doi.org/10.1007/978-3-030-86144-5_26

To conceive the proper place of informatics in the curriculum, it is natural to compare with subjects of similar character. One will then realise, that languages and mathematics are the closest analogies. Common for the three is also their character as tools for many other subjects.

Once informatics has become well established in general education, the mystery surrounding computers in many people's perceptions will vanish. This must be regarded as perhaps the most important reason for promoting the understanding of informatics. This is a necessary condition for humankind's supremacy over computers and for ensuring that their use do not become a matter for a small group of experts, but become a usual democratic matter, and thus through the democratic system will lie where it should, with all of us.

Naur's plea to properly include informatics in general education with a standing similar to languages and mathematics and his arguments for the plea are not less relevant today, more than half a century after their articulation.

In June 2018, *The Atlantic* brought the chronicle *How the Enlightenment Ends* by Henry Kissinger (2018); in here, the author writes:

Heretofore, the technological advance that most altered the course of modern history was the invention of the printing press in the 15th century, which allowed the search for empirical knowledge to supplant liturgical doctrine, and the Age of Reason to gradually supersede the Age of Religion. Individual insight and scientific knowledge replaced faith as the principal criterion of human consciousness. Information was stored and systematized in expanding libraries. The Age of Reason originated the thoughts and actions that shaped the contemporary world order.

But that order is now in upheaval amid a new, even more sweeping technological revolution whose consequences we have failed to fully reckon with, and whose culmination may be a world relying on machines powered by data and algorithms and ungoverned by ethical or philosophical norms.

Informatics in general, and the particular development of artificial intelligence (AI), is changing human knowledge, perception, and reality – and, in so doing, changing the course of human history. Informatics has made it possible to automate an extraordinary range of tasks and has done so by enabling machines to play a role – an increasingly decisive role – in drawing conclusions from data and then taking action. The growing transfer of judgment from human beings to machines denotes the revolutionary aspect of informatics and of AI as described by Kissinger in the aforementioned chronicle.

More than half a century after Naur's plea, informatics is finally becoming a school subject. Internationally there is a quite fast emerging breeze in the direction of making informatics part of national curricula and part of general education for all. This trend reflects the growing recognition that informatics is an important foundational competence along with "the three Rs": reading, writing, and arithmetic/mathematics.

However, informatics curricula in general tend to prioritize technical content (computing systems, networks and the Internet, data and analysis, algorithms and programming) – perhaps adding an element of impact of computing.

The Danish curriculum for general education (one for upper secondary and a more novel for primary and lower secondary) acknowledges the bipartite nature of all computations directed at purposes in the real world – the problem domain and the

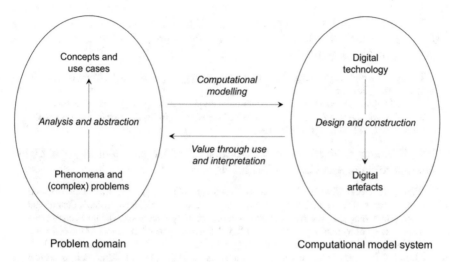

Fig. 1 The bipartite nature of computing

computational model system – as well as the relation between the two: representation and interpretation (see Fig. 1).

The equal inclusion of problem domain and interpretation, complementing computational model system and representation, is rather unique and embodies the Danish curriculum's perspective on digital humanism.

2 Political Emphasis on Informatics Education for All

In 2016, former president Obama launched *CS for All* as a bold new initiative to empower all American students from kindergarten through high school to learn computer science and be equipped with the computational thinking skills they need to be creators in the digital economy, not just consumers, and to be active citizens in our technology-driven world (White House 2016).

In 2018, major European and international organizations formed the coalition *Informatics for All* (Informatics for All 2018). In many ways, the Informatics for All initiative mirrors Obama's CS for All initiative. A crucial element of the European approach, which distinguishes it from the CS for All initiative, is the two-tier strategy at all educational levels: informatics as a discrete subject, that is, a fundamental and independent subject in school (like language and mathematics), and the integration and application of informatics with other school subjects, as well as with study programs in higher education. Perhaps overly simplified, the two tiers may be characterized as *Learn to Compute* (discrete subject) and *Compute to Learn* (integration); see Caspersen et al. (2019).

In 2020, the EU Commission launched Digital Education Action Plan for 2021–2027, which mentions informatics as a priority area and action (European Commission 2020a, p. 9):

> To support a high-performing digital education ecosystem, the European Commission will pursue the following actions: [...]
>   10. Propose a Council recommendation on improving the provision of digital skills in education and training [...] through a focus **on inclusive high-quality computing education (informatics) at all levels of education**.

The EU Commission highlights informatics education as a tool to boost digital competence (European Commission 2020b, p. 47):

> **Informatics education in school** allows young people to gain a critical and hands-on understanding of the digital world. If taught from the early stages, it can complement digital literacy interventions. The **benefits are societal** (young people should be creators not just passive users of technology), **economic** (digital skills are needed in sectors of the economy to drive growth and innovation) **and pedagogical** (computing, informatics and technology education is a vehicle for learning not just technical skills but key skills such as critical thinking, problem solving, collaboration and creativity).

In most European countries, the ambition is not yet as high as it ought to be considering the impact that the digital revolution has had on all aspects of society (CECE 2017; CECE's map 2017), and therefore the European Commission's emphasis on inclusive informatics education at all levels of education is both important, necessary, and timely.

3   The Danish Informatics Curriculum for General Education

Various flavors of informatics have been a topic in Danish upper secondary schools for more than 50 years (Caspersen and Nowack 2013).

In late 2008, the Ministry of Education established a task force to conduct an analysis of informatics in upper secondary schools and provide recommendations for a revitalization of the subject, not as a niche specialty but as a general subject relevant for all. Subsequently, a new general, coherent, and uniform informatics subject was developed, tested, and finally made permanent in 2016, however, not yet as a compulsory subject for all upper secondary education.

A distinct aspect of the Danish informatics curriculum is the focus on *digital empowerment*. We define digital empowerment as a concern for how students, as individuals and groups, develop the capacity to understand digital technology and its effect on their lives and society at large and their ability to engage critically and curiously with the construction and deconstruction of digital artifacts (Dindler et al. 2021).

An approach to embrace digital empowerment was present already in the Danish upper secondary informatics curriculum developed in 2009 (Caspersen 2009). One

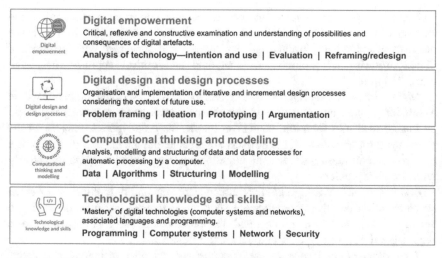

Fig. 2 The four competence areas in the Danish informatics curriculum for primary and lower secondary school

of the six key competence areas was *Use and impact of digital artifacts on human activity*. The purpose of this competence area was that students should understand that digital artifacts and their design have a profound impact on people, organizations, and social systems. Design of a system is not just design of the digital artifact and its interface, it is also design of the use and workflow that unfolds around the artifact. The purpose is that the students understand the interplay between the design of a digital artifact and the behavioral patterns that intentionally or unintentionally unfolds (Caspersen and Nowack 2013).

The informatics curriculum for primary and lower secondary education was developed by mandate of the Danish Ministry of Education in 2018 and is running on trial until 2021 in about 5% of primary and lower secondary schools across Denmark.

The author of this chapter and a colleague from the Department of Digital Design and Information Studies at Faculty of Arts were invited to serve as chairs for the group developing the curriculum. In choice of chairs, the Minister of Education signaled the importance of integrating a digital humanism perspective in the design of the curriculum.

The informatics curriculum for primary and lower secondary school consists of four competence areas (Danish Ministry of Education 2018):

- Digital empowerment
- Digital design and design processes
- Computational thinking and modeling
- Technological knowledge and skills

An overview of the four competence areas is provided in Fig. 2.

Digital empowerment refers to the critical and constructive exploration and analysis of how technology is imbued with values and intentions and how it shapes our lives as individuals, groups, and as a society. It is concerned with the ethics of digital artifacts and promotes an analytical and critical approach to digital transformation.

Digital design and design processes refers to the ability to frame problems within a complex problem area and, through iterative processes, generate new ideas that can be transformed into form and content in interactive prototypes. It focuses on the processes through which digital artifacts are created and the choices that designers have to make in these processes, highlighting students' ability to work reflectively with complex problems.

Computational thinking and modeling concerns the ability to translate a framed problem into a possible computational solution. It focuses on students' ability to analyze, model, and structure data and data processes in terms of abstract models (e.g., algorithms, data models, and interaction models).

Technological knowledge and skills concerns knowledge of computer systems, digital tools and associated languages, and programming. It focuses on the students' ability to express computational ideas and models in digital artifacts. This includes the ability to use computer systems and the language associated with these and to express ideas through programming. Working within this area aims at providing students with the experience and abilities needed to make informed choices about the use of digital tools and technologies.

Two of the competence areas – *Computational thinking and modeling* and *Technological knowledge and skills* – encompass classic computing subjects, while the two others are less standard in informatics curricula, if present at all.

4 Digital Humanism in Informatics: The Danish Perspective

The four competence areas constitute a holistic approach to informatics, here described in terms of the model in Fig. 1.

The real or an imaginary world is populated with phenomena and activities, which – through analysis and abstraction – can be understood in terms of concepts and use cases. Through computational modeling, these can be prioritized, structured, and modeled for computational representation. Through design and construction using digital technology, new digital artifacts can be developed to manipulate and transform these representations into something, which hopefully provides value through interpretation and use back in the real world (Madsen et al. 1993, chap. 18). The four competence areas map more or less one-to-one to the four processes in the model (Fig. 3).

Digital design and design processes covers primarily activities related to the problem domain. Computational thinking and modeling maps to the process going

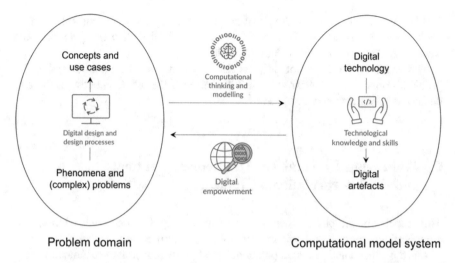

Fig. 3 Mapping of the four competence areas to the four processes in computational modeling of a problem domain

from problem domain to the computational model system. Technological knowledge and skills provides competences to design and construct digital artifacts. Digital empowerment relates to the process back from the computational model system to the problem domain.

Not only do the four competence areas constitute a nice, full circle. They also mutually strengthen each other. Analytic competences provided by digital empowerment will strengthen the three other competence areas. And vice versa: if you actually know how to build digital artifacts, your critical and constructive analysis can become much richer and deeper.

5 The Bipartite Nature of Informatics

The computational model system (computer systems, networks, security and programming languages, etc.) is a classic component for an informatics curriculum. No surprises here.

Inclusion of *Digital design and design processes* recognizes the bipartite nature of all computation that is directed at purposes in the real world. Thus, we embrace both problem domain and solution domain: the entire bipartite system – both the software machine and the physical (or imaginary) world whose behavior it governs.

This is not generally embraced in informatics curricula for general education. The focus on design process is inspired by the Scandinavian school of Participatory Design, which originated in the 1970s with subsequent development and proliferation beyond Scandinavia (Greenbeaum and Kyng 1991). It is also inspired by Donald Schön's philosophy of design as a reflective practice from the 1980s

(Schön 1983). The particular notion of problem framing and reframing are essential parts of this and is also inspired by the seminal work of British computer scientist Michael Jackson in the 1990s (Jackson 2000).

However, the focus is not only on the two parts of "the bipartite system" – problem domain and solution domain – but also on the relations between the two parts: *representation* and *interpretation*.

6 Embracing Uncertainty: The Representational and the Interpretational Challenge

Most aspects of the physical world, which we attempt to capture and represent in computational models and artifacts, are blurred, uncertain, and nondeterministic.

On the other hand, the computational models we construct are fundamentally strict, certain, and deterministic.

The challenge has two faces. One is *the representational challenge*: How can we model the blurred, uncertain, and nondeterministic aspects of the world in computational artifacts?

The other is *the interpretational challenge*: How do we avoid to constrain and eventually dehumanize our understanding of phenomena and concepts in the real world when our worldview is increasingly defined through the lenses of strict, certain, and deterministic computational models and artifacts?

The representational challenge is addressed by the competence area *Computational thinking and modeling* (data, algorithms, structuring, etc.), which is again a self-evident component in an informatics curriculum.

The interpretational challenge is addressed by the competence area *Digital empowerment*, which represents the ability to analyze and evaluate digital artifacts with a focus on intention and use through a critical, reflexive, and constructive examination and understanding of consequences and possibilities of a digital artifact. This competence area is for digital artifacts what literature analysis is for novels, but with the additional liberating component of reframing and redesign – realizing that digital artifacts are human-made and could have been designed differently if other perspectives had been applied.

7 Conclusions

We have presented elements of the current international development regarding informatics in general education for all, and we have presented the current development in Denmark in upper secondary school and more recently in primary and lower secondary school.

We have presented the distinct Danish perspective on digital humanism represented by an approach to informatics in school where the four competence areas of the curriculum constitute a holistic approach recognizing the bipartite nature of all computation by addressing problem domain as well as solution domain – but also the relations between the two: representation and interpretation.

In doing so, we aim at addressing two particular concerns related to computational artifacts:

1. How can we meaningful model the blurred, uncertain, and non-deterministic aspects of the world in computational artifacts?
2. How do we avoid to constrain and eventually dehumanize our understanding of phenomena and concepts in the real world when our worldview is increasingly defined through the lenses of strict, certain, and deterministic computational models and artifacts?

These are essential issues on our way forward into an in all aspects increasingly digital society. It is imperative that informatics become a fundamental and general discipline in school in order to ensure that future generations become educated and empowered to contribute to the development of their digital environment and to realize our technological advancement to ensure the evolution of a safe, secure, environmentally conscious and just society.

Acknowledgment I would like to thank the anonymous reviewers for valuable feedback to an earlier version of the manuscript.

References

Caspersen, M.E. (2009). Kernekompetencer i informationsteknologi (in Danish). Notes for ministerial working group. Accessed 21st April 2021.

Caspersen, M.E. & Nowack, P. (2013). Computational Thinking and Practice — A Generic Approach to Computing in Danish High Schools, Proceedings of the 15th Australasian Computing Education Conference, ACE 2013, Adelaide, South Australia, Australia, pp. 137-143.

Caspersen, M.E., Gal-Ezer, J., McGettrick, A.D. & Nardeli, E. (2019). Informatics as a Fundamental Discipline for the 21st Century. Communications of the ACM 62 (4), DOI:https://doi.org/10.1145/3310330.

CECE (2017). Informatics Education in Europe: Are We All In The Same Boat?, Report by the Committee on European Computing Education, Informatics Europe and ACM Europe. Accessed 12th March 2021.

CECE's Map (2017). CECE's Map of Informatics in European Schools, ACM Europe and Informatics Europe. Accessed 12th March 2021.

Danish Ministry of Education (2018). Indholdet i forsøgsfaget teknologiforståelse (in Danish). The Danish Ministry of Education. Accessed 14th March 2021.

Dindler, C., Iversen, O.S., Caspersen, M.E. & Smith, R.C. (2021). Computational Empowerment. In Computational Thinking Education in K-12: Artificial Intelligence Literacy and Physical Computing, MIT Press, 2021. Eds. Kong, S-C & Abelson, H. Scheduled for publishing in spring 2021.

Greenbeaum, J. & Kyng, M., Eds. (1991). Design at Work: Cooperative Design of Computer Systems. CRC Press.

European Commission (2020a). Digital Education Action Plan (2021-2027) – Resetting education and training for the digital age. European Commission. Accessed 12th March 2021.

European Commission (2020b). Commission Staff Working Document. European Commission. Accessed 12th March 2021.

Informatics for All (2018). Informatics for All. Accessed 12th March 2021.

Jackson, M. (2000). Problem Frames: Analyzing and structuring software development problems. Addison-Wesley.

Kissinger, H. (2018). How the Enlightenment Ends. The Atlantic, June 2018. Accessed 21st April 2021.

Madsen, O.L., Møller-Pedersen, B. & Nygaard, K. (1993). Object-Oriented Programming in the BETA Programming Language. Addison-Wesley.

Naur, P. (1967). Datalogi – læren om data (in Danish). The second of five Rosenkjær Lectures in Danish Broadcasting Corporation 1966-67 published as *Datamaskinerne og samfundet*, Munksgaard. Accessed 21st April 2021.

Naur. P. (1992). Computing: A Human Activity, ACM Press.

Schön D.A. (1983). The Reflective Practitioner: How professionals think in action. Temple Smith.

White House (2016). Computer Science For All. The White House. Accessed 12th March 2021.

The Unbearable Disembodiedness of Cognitive Machines

Enrico Nardelli

Abstract Digital systems make up nowadays the communication and social infrastructure and fill every parcel of space and time, affecting our lives both professionally and personally. However, these "cognitive machines" are completely detached from the human nature, whose comprehension is beyond their capabilities. It is therefore our duty to ensure their actions respect human rights and values of a democratic society. Education is one of the main tools to attain this goal, and a generalized preparation in the scientific basis of the digital technologies is a required element. Moreover, it is fundamental to understand why the digital automation has a nature completely different from the traditional industrial one and to develop an appreciation for human and social viewpoints in the development and deployment of digital systems. These are the key issues considered in this chapter.

This chapter discusses the theme of informatics education in the perspective of Digital Humanism. Given the breadth of impact on our society of the digital transformation, it is important to assume a historical viewpoint so as to better understand how to develop the most appropriate educational approach to accompany this epochal challenge.

Everybody is aware to live in the midst of a revolution. It is not what is usually identified as the Fourth Industrial Revolution (Schwab 2015), but what has been called the third "power revolution": this idea was first hinted at in Nardelli (2010), then sketched in Nardelli (2014, 2016) and discussed in more detail in Nardelli (2018), also in reference to education. The industrial society (which between the eighteenth and twentieth centuries replaced the previous agricultural society) is rapidly becoming an increasingly digital society. This revolution, made possible by the technology of the digital computer, and explained and fuelled by the scientific achievements of informatics, has allowed building, for the first time in the history of humanity, devices that have been defined "cognitive machines" in Nardelli (2018).

E. Nardelli (✉)
Università di Roma "Tor Vergata", Rome, Italy
e-mail: nardelli@mat.uniroma2.it

These are automatic systems that – by manipulating signs they ignore the meaning of, according to instructions they ignore the meaning of – transform data that have a sense for human beings, in a way that is significant to them.

The comprehension of this aspect is crucial to educate for Digital Humanism appropriately. Informatics can be aptly defined **the science of automated processing of representations** since it does not deal with "concrete" objects but with their representations, which are built by means of a set of characters taken from a finite alphabet. When human beings look at a representation, they usually tend to associate it to a meaning, which very often depends on the subject and is shaped by shared social and cultural conventions. For example, the sequence of characters "camera" will evoke one meaning to English speakers and a different one to Italians. Devices produced by the technological developments of informatics, instead, deal with representations without any comprehension of their meaning. Moreover, they process these representations, that is, they receive input representations and produce output representations (maybe through the production of a very long series of intermediate or ancillary representations) by acting in a pure automatic (or mechanical) way (Denning 2017; Nardelli 2019). Again, they do not have any comprehension of the meaning of the transformation processes they execute. However, in the end, these machines carry out operations that, considered from the viewpoint of human beings, are of a cognitive nature and meaningful to them.

These machines are now pervading the entire society, and this represents a real revolution, the "informatics revolution" – called the third "revolution of power relations" (Nardelli 2018), because for the first time in the history of humanity, cognitive functions are carried out by machines. This third revolution "breaks" the power of human intelligence, creating artifacts that can mechanically replicate cognitive actions, which until now were a characteristic of people.

Every scholar in our field is aware that these cognitive machines are characterized by a meaningless process of transformation (meaningless from the point of view of the mechanical subject carrying out the transformation) which produces a meaningful outcome (meaningful in the eyes of the human observer of the transformation). However, outsiders and common people usually neglect this. To properly understand the challenges education for Digital Humanism faces, a full and deep comprehension of this aspect is absolutely required. That is why it is important to discuss this revolution in the historical perspective of other equally disruptive revolutions in the history of humankind, to understand similarities and differences among them.

1 The Three Revolutions

Let us consider what happened previously: the first two revolutions in power relations, bringing radical changes both in the technical and in the social spheres, were the printing press one and the industrial one.

The invention of movable type printing in the fifteenth century caused both a technical and a social revolution in society: a technical one, because it made possible to produce texts in a faster and cheaper way, and a social one, because it made possible a more widespread circulation of knowledge. Ultimately, what happened was *the first revolution in power relations*: authority was no more bound to the spoken word; it was no longer necessary to be in a certain place at a certain time in order to know and learn from the living voice of the master. Knowledge always remains a great power, but now this power is not confined any more to the people who possess it or to those who are able to be close to them in time and space. The replicability of the text implies the replicability at a distance of time and space of the knowledge contained in it. All those who can read can now have access to knowledge. This set in motion epochal social changes: the diffusion of scientific, legal, and literary knowledge gave a huge impact to the evolution of society, which became increasingly democratic.

In fact, in the space of two and a half centuries, almost 800 million books printed in Europe caused an irreversible process of social evolution. Scientific knowledge, revolutionized by the Galilean method, thanks to the printing press, spread throughout Europe and constituted one of the enablers of the subsequent revolution, the industrial one, identified as *the second revolution in power relations*.

Started in the eighteenth century, this was equally disruptive: the availability of industrial machines made the physical work of people replicable. Human arms now are no longer needed because machines operate in their place. A technical revolution is achieved, because artifacts are replicated faster and in the absence of human beings. Machines can produce day and night without getting tired, and they can even produce other machines. They are amplifier and enhancer of the physical capabilities of human beings. A social revolution is obtained: physical limitations to movement and action are broken down. A single person can move huge quantities of dirt with a bulldozer, travel quickly with a car, and talk to anyone in the world through a telephone. Evolution and progress of human society are therefore further accelerated by the possibility of producing physical objects faster and more effectively, not to mention the consequences in terms of transporting people and things. The power relation that is revolutionized in this case is that between man and nature: humanity subjugates nature and overcomes its limits. One can quickly cross the seas, sail the skies, harness water and fire, and move mountains.

The printing press revolution had given humanity an extra gear on the immaterial level of information; the Industrial Revolution did the same for the material sphere. The world is filled with "physical artifacts" (the industrial machines) that begin to affect the nature of the planet in an extensive and deep way.

Then, in the middle of the twentieth century, after the realization of about 800 billion industrial machines,[1] *the third revolution in power relations*, that of information technology (IT), slowly begins. At first, it seems to be nothing more than an evolved variant of the automation of physical work and production processes caused by the Industrial Revolution, but after a few decades, we begin to understand

---

[1] Estimate by the author of the overall number of industrial machines of various kinds produced since around 1700 until more or less 1950.

that it is much more than that, because it affects the cognitive level and not the physical one. We are no longer replicating the static knowledge brought by spoken words and the physical strength of people and animals, but that "actionable knowledge" which is the real engine of development and progress. This expression denotes that kind of knowledge which is not just a static representation of facts and relationships but also a continuous processing of data exchanged dynamically and interactively between a subject and the surrounding context.

Because of the informatics revolution, this actionable knowledge (i.e., knowledge ready to be put into action) is reproduced and disseminated in the form of software programs, which can then be adapted, combined, and modified according to specific local needs. The nature of the artifacts, of the machines we produce, has changed. We no more have concrete machines, made by many physical substances: we now produce immaterial machines, made by abstract concepts, ultimately boiling down to configurations of zeroes and ones. We have the "digital machines," born – due to the seminal work of Alan Turing (Turing 1936) – as pure mathematical objects, capable of computing any function a person can compute, and which can be made concrete by physically representing them under some form, it does not matter which. Indeed, beyond the standard implementation of digital machines by using levels of voltage that, in some physical electric circuit, give substance to their abstract configurations, we also have purely mechanical implementations, with levers and gears, or hydraulic ones.

Even though these digital machines clearly require some physical substrate to be able to operate, they are no more physical artifacts. They are "dynamic cognitive artifacts," frozen action that is unlocked by its execution in a computer and generates knowledge as a result of that execution. Static knowledge of books becomes dynamic knowledge in programs. Knowledge capable of automatically producing new knowledge without human intervention. Therefore, most appropriately, they have been defined "cognitive machines" (Nardelli 2018).

2   Cognitive Machines

These machines are a reminiscence of those that, in the course of the Industrial Revolution, made possible the transformation from the agricultural society to the industrial one. Actually, they are different and much more powerful. Industrial machines are amplifiers of the physical strength of man; digital machines produced by the informatics revolution are cognitive machines (or "knowledge machines"), amplifiers and enhancers of people's cognitive functions. They are devices that boost the capabilities of that organ whose function is the distinctive trait of the human being.

On the one hand, we have a technical revolution, that is, faster data processing; on the other hand, we also have a social revolution, that is, the generation of new knowledge. In this scenario, what changes is the power relation between human intelligence and machines. Humankind has always been, throughout the history, the master of its machines. For the first time, this supremacy is challenged.

Cognitive activities that only humans, until recently, were able to perform are now within the reach of cognitive machines. They started with simple things, for example, sorting lists of names, but now they can recognize if a fruit is ripe or if a fabric has defects, just to cite a couple of examples enabled by that part of informatics that goes under the name of artificial intelligence. Certain cognitive activities are no longer the exclusive domain of human beings: it has already happened in a large set of chessboard games (checkers, chess, go, etc.), standard fields for measuring intelligence, where now the computer regularly beats the world champions. It is happening in many work activities that were once the exclusive prerogative of people and where now the so-called bots, computer-based systems based on artificial intelligence techniques, are widely used.

There are at least two issues, though, whose analysis is essential in the light of the educational viewpoint discussed in this chapter.

The first issue is that these cognitive machines have *neither the flexibility nor the adaptability to change their way of operating when the context they work in changes.* It is true that modern "machine learning"-based approaches give some possibility for them to "sense" changes in their environment and to "adapt" their actions. However, this adaptation space has its own limits. Designers must somehow have foreseen all possible future scenarios of changes, in a way or another. People are inherently capable of learning what they do not know (whereas cognitive machines can only learn what they were designed for) and have learned, through millions of years of evolution, to flexibly adapt to changes in the environment of unforeseen nature (while knowledge machines can – once again – only adapt to changes of foreseen types). We cannot therefore let them work on their own, unless they operate in contexts where we are completely sure that everything has been taken into account. Games are a paradigmatic example of these scenarios. These cognitive machines are automatic mechanisms, giant clocks that tend to behave more or less always in the same way (or within the designed guidelines for "learning" new behaviors). This is why digital transformation often fails: because people think that, once they have built a computer-based system, the work is completed. Instead, since no context is static and immutable, IT systems not accompanied by people able to adapt them to the evolution of operational scenarios are doomed to failure.

The second problem is that these cognitive machines are *completely detached from what it means to be human beings.* Someone can see it as a virtue, while on the contrary it is a huge flaw. There is no possibility of determining a single best way of making decisions. Those who think that algorithms can govern the human society in a way that is the best for everyone are deluded (or have hidden interests). Since the birth of the first forms of human society, the task of politics has been to find a synthesis between the conflicting needs that always exist in every group of people. Moreover, the production of this synthesis requires a full consideration of our human nature. The only intelligence that can make decisions appropriate to this context is the embodied intelligence of people, not the incorporeal artificial intelligence of cognitive machines. This does not imply there is not a role for cognitive machines.

Their use should remain confined to those of powerful personal assistants, relieving us from the more repetitive intellectual work, helping us in not making mistakes due to fatigue or oversight, and without leaking our personal data in the wild. People have always to remain in control, and the final decisions, above at all those affecting directly or indirectly other individuals and their relations, should always be taken by human beings. To discuss a current topic, it is understandable to think that, to some degrees, final decision of judges in routine cases may be affected by extrajudicial elements, unconscious bias and contingent situations and emotions. After all, even well-trained judges are anyhow fallible human beings. However, speculating on the basis of data correlation, as done in Danziger et al. (2011), that after lunch judges tend to be more benevolent is wrong. A more careful analysis highlighted other organizational causes for this effect (Weinshall-Margel and Shapard 2011). A cognitive machine just learning from data without a thorough understanding of the entire process would have completely missed the point. That is why the decision in France to forbid analytics on judges' activity is well taken (Artificial Lawyer 2019). Because incorporeal decision systems convert a partial description of what happened in the past in a rigid prescription of how to behave in the future, stealing human beings of their most precious and more characteristic qualities, free will.

Cognitive machines are certainly useful for the development of human society. They will spread more and more, changing the kind of work executed by people. This has already happened in the past: in the nineteenth century, more than 90% of the workforce was employed in agriculture; now it is less than 10%. It is therefore of the utmost importance that each person is appropriately educated and trained in the conceptual foundations of the scientific discipline that makes possible the construction of such cognitive machines. Only thus, everyone will be able to understand the difference between what they can do and what they cannot and should not do.

3   A Broader Educational Horizon

Education on the principles of informatics should start since the early years in school (Caspersen et al. 2019). The key vision that a digital computing system operates without any comprehension, by the system itself, of what is processed and how it is processed, needs to accompany the entire education process. Moreover, it should always go hand in hand with the reflection that the process of modeling reality in terms of digital data and processing them by means of algorithms is a human activity. As such, it may be affected by prejudice and ignorance, both of whom may, at times, be unconscious or unknown.

Only in such a way, in fact, it will be possible to understand that *any choice*, since the very first ones regarding which elements to represent and how to represent them, to the ones deciding the rules for the processing itself, *is the result of a human decision process and is therefore devoid of the absolute objectivity that too often is associated to algorithmic decision processes.*

Moreover, as Giuseppe Longo has observed, the fundamental distinction introduced by Turing between hardware and software is a "computational folly" when applied to living entities and society (Longo 2018). Firstly, because in the biological world there is not such a distinction between hardware and software. DNA, the code of life, constitutes its own hardware. The rewriting of digital representations happening in digital machines is different from the transcription from DNA to RNA. Secondly, because fluctuations are completely absent in the discrete world where Turing machines operate, while they play an essential role in complex dynamical systems all around us. As first noted by Henri Poincaré, this possibly results in the unpredictability of their evolution, even though they are deterministically defined (Poincaré 1892). Thirdly, because every software is only able to represent an abstraction of a real phenomenon. While this abstraction can provide valuable indications regarding its dynamics, considering the representation as the phenomenon itself – as it happens with scientists like Stephen Wolfram, who says: "The Universe is a huge Turing machine" (Wolfram 2013) – is as wrong as mistaking the map for the territory. And, finally, digital systems, once set in the same starting conditions within a same context, will identically compute always the same result, even for those complex systems where (as Poincaré proved) this is physically absurd. "Computer networks and databases, if considered as an ultimate tool for knowledge or as an image of the world" writes Longo "live in the nightmare of exact knowledge by pure counting, of unshakable certainty by exact iteration, and of a 'final solution' of all scientific problems."

To complete this discussion, note that the most recent developments in the theory of computation (van Leeuwen and Wiedermann 2012) have shown that models more powerful than the Turing machine[2] are needed to model the interactivity with external agents, the evolution of the machine's initial program, and the unbounded operations over time, three aspects that characterize modern computing systems. Note, though, that while these approaches offer more appropriate formal models to describe what happens in the modern world of continuously running digital devices interacting with people, they require the introduction of components that are Turing-uncomputable. Therefore, understanding the difference between the (mechanical) role played by digital computing systems and the (uncomputable) behavior exhibited by human beings *is essential to be able to use cognitive machines to increase the well-being of human beings and not to oppress them.*

On the other side, we have moved a large part of our life in the realm where these disembodied cognitive machines rule. Consequently, our existence now develops not only along the usual relational dimensions (economical, juridical, cultural, etc.) but articulates also in this incorporeal dimension of "representations" that is more and more relevant, from a social point of view. Humanity has been recording data about the world for thousands if not tens of thousands of years. However, from being a completely negligible component of our existence, representations of data have

---

[2]These are the so-called red-green Turing machines. See van Leeuwen and Wiedermann (2012) for their description and a list of other equivalent models.

become a relevant and important part of it. As the health emergency of 2020 has unfortunately taught us, we cannot disregard them any longer. They have become an integral and constitutive component of our personal and social life. Hence, the necessity of giving protection to people rights not only for what regards their body and their spirit but also to their digital projections (Nardelli 2020).

As a side note, note that the disembodiedness of cognitive machines has a dual counterpart in the fact that this digital dimension of our existence is populated by "life forms" which we have not the sensor to be aware of. Digital viruses and worms, which are not benign towards our "digital self," much like their biological counterparts are not benevolent to our physical bodies, continue to spread at an alarming rate without us being able to counter them effectively. Indeed, we would need the digital counterpart of those hygiene rules that such a big role have had in the improvement of living conditions in the twentieth century (Corradini and Nardelli 2017). Once again, it is only through education that we can make the difference, and it has to start as early as possible.

While the general education of citizens happening in school should be focused on the above principles, when considering the tertiary education level, something more is needed.

We need to prepare our students in a way similar to how we train medical doctors. In the early years, they study the scientific basis of their field: physics, chemistry, and biology. In this context, universal and deterministic laws apply. Then, as they progress in their educational path, aspiring doctors begin the study of the "systems" that make up human beings (motor, nervous, circulatory, etc.), thus learning to temper and combine mathematical determinism with empirical evidence. Finally, when they "enter the field," they face the complexity of a human being in its entirety, for whom, as general practitioners well know, a symptom can be the manifestation not only of a specific disease but also of a more general imbalance. At this point, the physician will no longer be able to apply simply one of those universal laws she learned in the early years. This does not mean abdicating the scientific foundations to return to magical rites or apotropaic formulas but acting to solve the specific problem of the specific patient she is facing, in the light of the science that she has introjected. The informatician, like the doctor, must have his feet firmly planted in science, but his head clearly aimed at making people and society feel better (just as a doctor does).

More specifically, informatics students should be prepared to have good basis in mathematics, algorithmic, semantics, systems, and networks, but then they should be able to solve automation problems regarding data processing (intended in its widest meaning) without making people appendices to IT systems. To support the goals of Digital Humanism, they should merge their "engineering" design capabilities with attention to human-centered needs. They should be educated to develop an appreciation for human and social viewpoints regarding digital systems. They have to tackle the challenges of digital transformation while improving social well-being of people and not only enriching the "owner of steam," who has every right to an adequate remuneration of his capital but not at the price of dehumanizing digital systems end users.

That is why we need to broaden the educational horizon of our study courses, complementing the traditional areas of study with interdisciplinary and multidisciplinary education, coming above all from the humanistic and social areas. Only in such a way it will be possible to recover the holistic vision of technological scenarios that is characteristic of a humanism-based approach, where respect for people and values of a democratic society are the guiding forces.

Acknowledgments The author thanks the anonymous reviewers for their insightful comments, which enabled to improve the presentation of this chapter.

References

Caspersen, M.E., Gal-Ezer, J., McGettrick, A., and Nardelli, E. (2019) 'Informatics as a fundamental discipline for the 21st century'. Communication of the ACM, 62(4).

Corradini, I. and Nardelli, E. (2017) *Digital hygiene: basic rules of prevention.* Link&Think, [Online]. https://link-and-think.blogspot.com/2017/11/digital-hygiene-basic-rules-of.html

Danziger, S., Levav, J., and Avnaim-Pesso, L. (2011) 'Extraneous factors in judicial decisions', Proceedings of the National Academy of Sciences, 108 (17).

Denning, P.J. (2017) 'Remaining trouble spots with computational thinking', Communication of the ACM, 60(6).

France Bans Judge Analytics, 5 Years In Prison For Rule Breakers, (2019) Artificial Lawyer, [Online]. https://www.artificiallawyer.com/2019/06/04/france-bans-judge-analytics-5-years-in-prison-for-rule-breakers/

van Leeuwen, J. and Wiedermann, J. (2012) 'Computation as an unbounded process', Theoretical Computer Science, 429:202-212, 2012.

Longo, G. (2018) 'Letter to Alan Turing', *Theory, Culture & Society*, Special Issue on "Transversal Posthumanities" (M. Fuller, R.Braidotti, eds.), 2018

Nardelli, E. (2010) *The maintenance is the implementation OR Why people misunderstand IT systems*, 6th European Computer Science Summit, [Online]. https://www.informatics-europe.org/ecss/about/past-summits/ecss-2010/conference-program.html

Nardelli, E. (2014) *Senza la cultura informatica non bastano le tecnologie*, (in Italian) [Online]. https://www.ilfattoquotidiano.it/2014/02/01/senza-la-cultura-informatica-non-bastano-le-tecnologie

Nardelli, E. (2016) *La RAI che vorrei: diffondere la "conoscenza in azione" per far crescere l'Italia*, (in Italian), Key4Biz, [Online]. https://www.key4biz.it/la-rai-che-vorrei-e-nardelli-contribuisca-allalfabetizzazione-digitale/156540/

Nardelli, E. (2018) *The third "power revolution"*, Link & Think, [Online]. https://link-and-think.blogspot.com/2018/05/informatics-third-power-revolution.html

Nardelli, E. (2019) 'Do we really need computational thinking?', Communication of the ACM, 62 (2).

Nardelli, E. (2020) *Does our "digital double" need constitutional rights?* Link&Think, [Online]. https://link-and-think.blogspot.com/2020/12/does-our-digital-double-need-constitutional-rights.html

Poincaré, H. (1892) *Méthodes Nouvelles*. (in French), Paris, 1892.

Schwab, K. (2015) *The Fourth Industrial Revolution*. Foreign Affairs, [Online]. https://www.foreignaffairs.com/articles/2015-12-12/fourth-industrial-revolution

Turing, A. (1936) 'On computable Numbers, with an application to the Entscheidungsproblem'. *Proc. London Mathematical* Society, 42, 1936.

Weinshall-Margel, K. and Shapard, J. (2011) 'Overlooked factors in the analysis of parole decisions', *Proceedings of the National Academy of Science*, 108 (42), Oct 2011.

Wolfram, S. (2013) 'The Importance of Universal Computation'. In *Alain Turing, his work and impact* (B. Cooper, ed.), Elsevier, Amsterdam, 2013.

Part VIII
Digital Geopolitics and Sovereignty

The Technological Construction of Sovereignty

Paul Timmers

Abstract For policy-makers, it has always been a struggle to do justice to a diversity of perspectives when tackling challenging issues such as market access regulation, public investment in R&D, long-term unemployment, etc. In this struggle, technology, as a force that shapes economy, society, and democracy, at best used to be considered as an exogenous factor and at worst was simply forgotten. Today, however, we live in a different world. Technology is recognized as a major driver. Digital technology is now in the veins, heart, and brains of our society. Yet, the idea that we can put technology to our hand to shape reality, rather than taking technology as a given, has still not been embraced by policy-makers. This chapter argues that we can and should give a stronger steer on technology to construct the kind of reality and in particular the kind of sovereignty we aspire.

1 Code Is Law; Law Is Code

Around the year 2000, Lawrence Lessig, a Harvard law professor, put forward his famous statement "code is law" (Lessig 2000). In brief, at that time, this was about the observation that the way the internet is technologically constructed ("code" in the sense of software code) to a large extent determines the rules of behavior in the internet. Code acts like law.

Recently, however, we would rather say: "law is code." "Law" is here understood as the requirements that governments would like to impose on the digital world. Nowadays, these requirements are evermore driven by concerns about sovereignty. Governments want more control over cybersecurity in 5G and open up access to

P. Timmers (✉)
University of Oxford, Oxford, UK

European University Cyprus, Engomi, Cyprus
e-mail: paul.timmers@iivii.eu

H. Werthner et al. (eds.), *Perspectives on Digital Humanism*,
https://doi.org/10.1007/978-3-030-86144-5_28

Fig. 1 Social and
technological construction
of reality

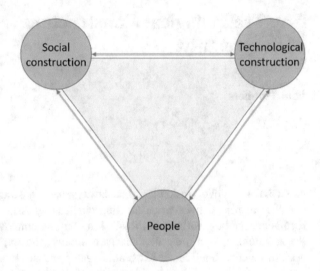

gatekeeper digital platforms.[1] States feel that they have to act to protect their national economic ecosystem and are worried about the erosion of society's values such as privacy. They fear that the very authority of government is being undermined. Clearly, technology as given does not safeguard sovereignty and has even become a threat. Sovereignty and strategic autonomy have become *Chefsache*.[2]

2 Social and Technological Construction of Reality

What is happening here from a conceptual point of view? I will stress two ideas, without claiming any originality in doing so.[3] The first is that technological construction of reality is as valid a notion as is the social construction of reality. The second is that there is a strong interplay between social and technological construction (Fig. 1).

The corollary is that design of social constructs such as law and design of technological constructs can and may go hand in hand. Even stronger: by ignoring that interplay, exploitative powers (dictators, populists, criminals, unscrupulous companies) will step in the void and gamble with our economies, societies, and democracy.

---

[1] For example, as in the EU reflected in the 5G Cybersecurity Recommendation and in the Digital Markets Act.

[2] GAIA-X, the European cloud initiative, takes (data) sovereign by design as a guiding principle for the development of software and services; see Franco-German Position on GAIA-X, 18 Feb 2020.

[3] For the origins of the underlying idea of constructivism, see Immanuel Kant, *Critique of Pure Reason* (1781).

The idea of social construction of reality rose to prominence from 1966 onward, thanks to Peter Berger and Thomas Luckmann (1967). Since that time, we accept that much of what we consider real in everyday life, such as money, values, citizens, or state, is a social construct. This holds for state sovereignty as well. Indeed, 30 years after Berger and Luckmann's *The Social Construction of Reality*, the excellent book *State Sovereignty as a Social Construct* was published (Biersteker and Weber 1996).

Can reality also be technologically constructed?[4] Pretty obviously "yes" when we just consider the many technological physical artifacts around us. These are the tangible technological reality "as we know it."

Such technological reality can even shape artifacts in our mind such as our perception of reality. Jean Baudrillard, a French sociologist and philosopher, argued in his provocative set of essays "The Gulf War Did Not Take Place" that this war was presented to us through technology with a specific imagery (Baudrillard 1991). Remember the cockpit images of bombing targets in the cross-hairs? These became for many the reality of the Gulf War (as long as you were not on the ground...). Technology-generated perception becomes evermore part of reality. Some young people have an unhealthy obsession with their image on social media (McCrory et al. 2020).

But can social reality, social artifacts, also be technologically constructed? The answer is affirmative here too. Consider Lessig's "code is law" as mentioned before. Lessig focused on the interplay of technology and law. Law is of course a social construct *par excellence*. Julie Cohen, in her 2017 book *Between Truth and Power*, built on Lessig and the 1970s governmentality concept of Michel Foucault (Cohen 2019). She analyzed the interplay of technological and social construction in the governance of law development by governments and tech companies. One conclusion: technology may be malleable, but such social constructs are malleable as well.

3 Technological Construction of Sovereignty

What then is technological construction of sovereignty? Sovereignty is loosely speaking about territory and borders, people, "our" values, and resources that "belong to us." Sovereignty requires internal legitimacy of the authority toward the people. Sovereignty also requires external legitimacy, that is, recognition by other states (Biersteker 2012; Timmers 2019).

Firstly, sovereignty includes assets that "belong to us." These are not only land or rivers but increasingly also technologically constructed assets, notably digital ones such as our health data or the country's internet domain name.

---

[4]That is, the reality of technological artifacts, technology mediating reality, and technology shaping or conditioning social reality.

Secondly, technology can redefine core privileges of the state such as the identification of citizens (the French call it *une fonction régalienne*). Electronic identity or eID raises the question of control. Can only a government-issued identity be an official eID? Could it also be a self-sovereign identity? Or even an identity owned by a platform like Facebook or Google? Should the *fonction régalienne* loose its state anchor? The technological choice, in combination with social constructs such as law and market power, can redefine a core aspect of sovereignty.

Thirdly, technology, in its Baudrillard's sense of intermediator to reality, unlocks cultural heritage which is clearly a sovereign asset. Technology, properly designed, protects and strengthens our values. Privacy by design is an illustration.

What then about digital technologies shaping internal and external legitimacy, those core qualities of sovereignty? Internal legitimacy implies accountability and transparency of the legitimate authority. As citizen we may wonder: is my court case treated fairly? Why have I been singled out for income tax scrutiny? Which civil servant is looking at my personal data?

On the one hand, transparency can be enabled by an appropriate technology architecture. Estonia has chosen to base its e-government platform on blockchain—which cannot be tampered with—for that purpose. On the other hand, internal legitimacy can also be undermined by technology that intentionally or unwittingly does not respect fundamental and human rights. In the Netherlands, recently "smart" but discriminatory technology for detecting misuse of child support in combination with strict bureaucracy and blind politics led to serious injustice for thousands of citizens. The Dutch government fell over the case. It lost its internal legitimacy.

The counterpart of technology-defined control of government is technology-defined control of citizens. Already today, even in free societies, ever-smarter cameras are ubiquitous. COVID-19 apps have raised concerns about surveillance creep (Harari 2020). Democratic processes everywhere are heavily shaped by social media, which stimulate by their very design the formation of echo chambers and thereby give raise to polarization. Hostile states seek to undermine the very legitimacy of incumbent governments by making use of the architecture of social media platforms to spread misinformation. Alternatively, social media are put under government control in order to suppress any citizen movement that may contest the state. This is a main motivation of online censorship in China (King et al. 2014).

External legitimacy can equally be shaped by technology. Kings and castles have fallen at the hand of new technologies such as trebuchet and cannon balls. The nuclear bomb prompted France to develop its own atomic capacity to safeguard its sovereignty. Asserting legitimacy in cyberspace has become a technological war where the power of one nation vis-à-vis others is increasingly being defined by militarizing artificial intelligence. One may wonder, though, what the nature is of such AI. How will it interpret aggression, and will it counterstrike autonomously or not? AI is a technology that can take over agency from the state, shaping external and internal legitimacy and thereby redefining sovereignty in the digital age.

Technological construction starts to reshape social constructs such as sovereignty. The writing is on the wall. The rise of cryptocurrencies challenges central banks as a sovereign institution. The rise of interoperable data spaces worries some data holders who fear that their autonomy is threatened. Data hoarding by digital platforms makes governments realize that their presumed sovereignty is evermore in the hands of a few global corporates and foreign governments. Technology is re-allocating legitimacy between the extremes of massive decentralization—such as with blockchain or personal data pods—and massive centralization in the hands of a few actors that escape democratic control.

4   Conclusions

The reader, having come all the way until here to read about something she or he already knew or at least intuited, may be left with the question: "so what?" The answer is that technology fundamentally shapes sovereignty and it is us who can influence the shaping of such technology.

Policy-makers who are concerned about strategic autonomy do not need to accept technology "as is." Technology is neither a force of nature nor to just be left to the market nor to be taken for granted, as an exogenous factor. Policy-makers can insist that (digital) technology is designed in such a way that internal and external legitimacies are strengthened. Digital technology can be required to be designed such as to grow assets "that belong to us" and protect our values, human rights, and humanism (TU Wien 2019).

Sure, we then sacrifice one holy cow as we must conclude that technology is not neutral. Fine. But there is a more radical proposition here, namely, that during the design of law policy-makers would sit together with technology designers. They would engage in a dialogue about technology requirements such as sovereignty safeguards. They would not be satisfied until there is mutual re-assurance of the compatibility of technology and law (or policy).

There is also no need to take the law and organization of government and administration for a given. Sure, we want stability with law. But if technology can do a better job, law-as-is should not stand in the way. This then leads to a second radical proposition: to consider in the design of any law whether promotion of technological disruption should be included in that same law. The intent would be to enable replacement of the social constructs in that law by technological constructs. Of course, only provided the end result is better.

An example would be to include in future laws that seek to safeguard sovereignty (such as on data or AI or cloud) a chapter on R&D for sovereignty-respecting technology, with corresponding budget and objectives. That same law should then foresee to scale back human oversight following proper and successful assessment of the resulting technology.

Co-design of law and technology in the way proposed here is not yet found anywhere, as far as the author is aware of. It would likely be seen as a radical change. But hopefully this chapter convinced the reader that this change is thinkable, enlightening, and, above all, necessary today in order to construct the sovereignty that we aspire. We have a choice.

References

Baudrillard, J. (1991) *La Guerre du Golfe n'a pas eu lieu*. Paris: Galilée.

Berger, P. and Luckmann, T. (1967) *The Social Construction of Reality*. Anchor.

Biersteker, T. J. (2012) 'State, Sovereignty and Territory', in Carlsnaes, W. (et al.) (ed.) *Handbook of International Relations*. SAGE Publications Ltd.

Biersteker, T. J. and Weber, C. (1996) *State Sovereignty as Social Construct, Cambridge Studies in International Relations*. Cambridge: Cambridge University Press. DOI: https://doi.org/10.1017/CBO9780511598685.

Cohen J. (2019) *Between Truth and Power: The Legal Constructions of Informational Capitalism*. Oxford University Press.

Harari Y. (2020) 'The world after coronavirus', *Financial Times*. Available at: https://www.ft.com/content/19d90308-6858-11ea-a3c9-1fe6fedcca75.

King, G., Pan, J. and Roberts, M. E. (2014) 'Reverse-engineering censorship in China: Randomized experimentation and participant observation', *Science*, 345(6199), p. 1251722. doi: https://doi.org/10.1126/science.1251722.

Lessig L. (2000) 'Code Is Law', *Harvard Magazine*, (1 Jan 2000). Available at: https://www.harvardmagazine.com/2000/01/code-is-law-html.

McCrory, A., Best, P. and Maddock, A. (2020) 'The relationship between highly visual social media and young people's mental health: A scoping review', *Children and Youth Services Review*, 115, p. 105053. doi: https://doi.org/10.1016/j.childyouth.2020.105053.

Timmers, P. (2019) 'Strategic Autonomy and Cybersecurity', *EU Cyber Direct*, (September 2017), pp. 1–10.

TU Wien (2019) *Vienna Manifesto on Digital Humanism*. Available at: https://dighum.ec.tuwien.ac.at/dighum-manifesto/.

A Crucial Decade for European Digital Sovereignty

George Metakides

Abstract The current decade will be critical for Europe's aspiration to attain and maintain digital sovereignty so as to effectively protect and promote its humanistic values in the evolving digital ecosystem. Digital sovereignty in the current geopolitical context remains a fluid concept as it must rely on a balanced strategic interdependence with the USA, China, and other global actors. The developing strategy for achieving this relies on the coordinated use of three basic instruments, investment, regulation, and completion of the digital internal market. Investment, in addition to the multiannual financial framework (2021–2027) instruments, will draw upon the 20% of the 750 billion recovery fund. Regulation, in addition to the Digital Governance Act and the Digital Market Act, will include the Data Act, the new AI regulation, and more that is in the pipeline, leveraging the so-called Brussels effect. Of key importance for the success of this effort remains the timing and "dovetailing" of the particular actions taken.

It looks increasingly likely that the decade of the 2020s will be crucial for the future of the economic, political, and social impact of digital technologies worldwide.

It will be particularly critical for Europe's aspiration to attain and maintain digital sovereignty while protecting and promoting cherished values that humanity honed and filtered through fifth century BC Athens and the eighteenth century AD Enlightenment to our days. Digital sovereignty remains a fluid concept as analyzed in Moerel and Timmers (2021). Here, it is not intended to mean digital autarky or absolute autonomy but rather a strategic positioning that ensures a globally balanced interdependence where a contestant's attempt to damage others risks self-damage.

At the beginning of the last decade, the 2010s, the digital landscape looked decidedly . . .rosier. It was still a time of "digital innocence" and great optimism for how, on balance, digital technologies would transform the world.

G. Metakides (✉)
Digital Enlightenment Forum, Zoetermeer, The Netherlands
e-mail: george@metakides.net

H. Werthner et al. (eds.), *Perspectives on Digital Humanism*,
https://doi.org/10.1007/978-3-030-86144-5_29

There was the Arab Spring of 2011 and the credible promise that digital technologies in general and the internet and its "social machines" in particular would help to create an informed citizenry that would, in turn, foster and strengthen democracy and social cohesion.

Then came the Snowden revelations and the Cambridge Analytica scandal in 2013, the Brexit referendum and the US elections in 2016, and an avalanche of related developments so that now, at the start of 2021, we are looking at misinformation, conspiracy theories, and fake news, turbo-charged by the social media platforms, having circled the globe many times over by the time truth is checked and outed. The 2011 Arab Spring case is now studied as a "Precursor of the Disinformation Age"!

We are also looking at fast-evolving, data-gobbling, non-transparent AI algorithms which, combined with the business models used by the social media platforms, push people to the extremes of the political spectrum, thus emptying the Aristotelian "middle" which is crucial for both democracy and social cohesion.

The rosy digital landscape, and its geopolitical context of the start of the decade of the 2010s, has been replaced at the start of the 2020s with a very different one where all kinds of red lights are flashing and warning bells are ringing.

The era of digital innocence and unbridled hope is over, replaced by mitigated belief that the accelerating advances in digital technologies can still be a force for good overall but only provided that governments, industry, academia, and individual citizens take actions that can ensure that this belief is realized.

Most of these actions currently proposed and debated are controversial and viewed very differently in the USA, China, Europe, and other parts of the world.

This is at the heart of a complex developing global power play in the context of which Europe is trying to position herself.

Before going further to see how this new decade might play out and why it might well be critical in shaping perhaps several decades that will follow, let us first take a look at the evolving geopolitical context.

Practically all recent studies conclude the same world superpower array as does the February 2021 Gallup report (Gallup International 2021) that polled perceptions on this very issue worldwide.

That is, by 2030, there will be two major superpowers combining economic and military prowess that will be jockeying for first place, the USA and China.

Quite far behind (further than now) will be the EU closely followed by Russia.

The falling further behind of the EU does not come as a surprise as the EU is not expected to come close to the two big ones without a common foreign and defense policy which, in turn, is not expected to happen during this decade.

But there is an additional finding in the aforementioned Gallup report. The second key question asked (besides which will be the major superpowers by 2030) asked which superpowers would be "stabilizing factors" and which would be destabilizing ones in the world (Fig. 1).

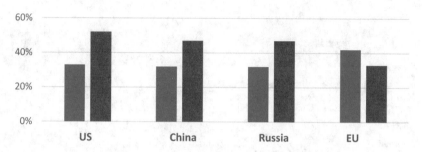

Fig. 1 Response to stabilizing and destabilizing powers

The EU is viewed as a stabilizing factor by a higher percentage of people around the globe than either of the two big ones or Russia.

Analysts concur that this perception is strongly correlated with a more general worldwide view of the EU as a non-aggressive soft power.

This also helps explain why EU regulatory initiatives like the GDPR which, despite any putative or real shortcomings, has brought data protection into the public eye have been rather quickly emulated worldwide by more than 120 countries and therefore why the EU can punch above its geopolitical weight as a soft power with global impact on regulatory issues.

All this has been thoroughly analyzed in Anu Bradford's excellent book *The Brussels Effect* (Bradford 2020) showing that European regulation of the digital landscape has the potential of becoming a blueprint for similar regulation around the world in areas such as data privacy, consumer health and safety, anti-trust, and online hate speech.

This, in turn, allows the EU to promote worldwide a European way as an alternative to the libertarian and authoritarian way of shaping the digital ecosystem currently followed in the USA and China, respectively.

Take the Big Tech oligarchy as an illustration.

It was during the decade of the 2010s that the GAFAM and their Chinese counterparts have attained their full dominance.

Now, at the start of 2021, we have GAFAM with a 7.6 trillion market value, strengthened during the pandemic and projecting a doubling in sales by the end of the decade.

The economic strength of the giants encourages their bulimic appetite for swallowing small innovators either to integrate them or to bury them. Apple, e.g., buys a company every 3 or 4 weeks and has bought over a 100 in the last 5 years.

First, winner takes all and then size begets size and power begets power.

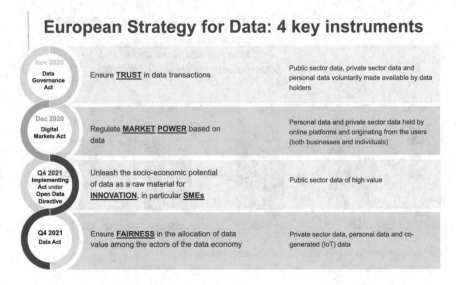

Fig. 2 Presentation of Dr. Yvo Volman at the Digital Humanism panel discussion of February 23, 2021 (DigHumTUWien 2021)

It was the EU that led the way to create a regulatory framework that would level the playing field and create a more contested digital economy, while the USA relied on voluntary self-regulation and China on her authoritarian power.

During the last decade, the USA has been, to put it mildly, skeptical about such regulation at the time as risking the stifling of innovation, but policies are now changing across the Atlantic as well.

Anti-competition issues including the possibility of break-ups are now on the table in the USA, and broader regulatory issues are proposed. Decisions taken (or not taken) there will have crucial impact on how the entire digital ecosystem evolves during the decade of the 2020s.

The EU and the USA remain at odds on a number of digital issues concerning citizen privacy, the related issue of date transfer across the Atlantic, and the ongoing spat of how to tax the tech oligarchs.

At the same time, especially after the 2020 US elections, both the USA and Europe are developing a common concern about China's use of digital technologies unhindered by ethical, privacy, and human rights concerns and the comparative advantage this could give China in leading the race to more and more advanced, data-dependent, AI technologies.

Following the example of GDPR, the EU has proposed or is about to propose an impressive, in scope at least, arsenal of regulatory measures including the AI regulation, Digital Services Act, Digital Market Act, Data Governance Act, and Data Act (see also Fig. 2).

Regulatory global influence constitutes for Europe a necessary, though by no means sufficient, step toward attaining and maintaining digital sovereignty.

Regulation complemented with investments and a completed digital internal market however does offer a more convincing strategy.

Massive investment in digital infrastructure and technology during this decade is now visible. The 150 billion euros comprising the 20% of the pandemic-triggered 750 billion recovery fund is expected to leverage much greater investments via the European Investment Bank and other institutions as well as both public and private investments at the national level. As an illustration, a European-federated, next-generation cloud service, building on the experience of GAIA-X, would be a top investment priority during the first half of the current decade. Additional planned investments would go toward doubling the EU share of global semiconductor production by 2030, universal EU 5G coverage, and quantum computing.

In parallel, completion of the European digital internal market would enable European digital companies to look at continent-wide opportunities with 450 million potential clients as a first step toward becoming competitive global players.

It is now recognized at the highest political decision-making level that European Digital Sovereignty is "central to European Sovereignty" and is "the capability that Europe must have to make its own choices, based on its own values, respecting its own rules" (quotes by the Presidents of the Council and the Commission, respectively).

At the same time, European Digital Sovereignty is necessary for empowering Europe to help ensure a digital transformation which is inspired, as stated in the Treaty on European Union, by "the humanist inheritance of Europe, from which we have developed the universal values of the inviolable and inalienable rights of the human person, freedom, democracy, equality and the rule of law" (Ministerial Council of the EU 2020).

During the current decade, every step taken toward European Digital Sovereignty will be unavoidably interacting with the EU's geopolitical positioning with respect to other global players and the positioning of China and the USA in particular.

Just as an illustration, China announced in early March 2021 the launch of a 5-year program aimed to decrease China's technological dependency on the West in 30 areas deemed of specific importance such as microprocessors, O.S. for mobiles, and aircraft design software. This comes, to a large extent, as a reaction to the steps taken by the USA in the last 4 years to curtail the power of Chinese tech giants like Huawei and SMI.

A week before the Chinese move, the Biden government in the USA has announced steps to decrease the dependency of the USA on rare earths and other imports affecting the digital supply chain.

Europe's ability to attain and maintain digital sovereignty during such continuous rebalancing of strategic interdependence among the global players expected during the 2020s will depend on how quickly and effectively it acts and reacts in such a dynamic geopolitical context.

In some cases, selective European autarchy may be the right move, while in others trusted alliances may be the best option. There may also be cases where planning to leapfrog to the next generation of technology may be preferable rather than trying to catch up on applications of current ones.

Quick decisions and follow-up steps are not easy for the European Union, an incomplete construct of 27 small, in the global context, countries that do not have a common foreign and defense policy which can provide a tool to help face China's state-supported industrial policy and the defense procurement innovation pull of the USA.

This presents a real challenge for the European Sovereignty during this decade.

To counter this inherent handicap, in some strategically selected cases, a subset of some EU countries could potentially move forward together quickly in a particular technology area where they are very advanced (e.g., quantum computing) with the possibility left open that other fellow member states could join later.

Perhaps selected quick, decisive moves during this decade in Europe could be more effective than the unavoidably slow process of attempting to tease out a full industrial policy equivalent first.

A truly unhindered digital internal market, regulation, and investments are actions that constitute powerful building blocks of a potentially effective Digital Sovereignty strategy for Europe.

Part of this strategy is the timing of new regulation and new investment in a particular area so that this investment and the players involved can draw maximum benefit as the regulation's "first users."

This is why the timeline of impact, the dynamic dovetailing of the specific actions emanating from aforementioned building blocks so as to maximize the added value resulting from combining them in a timely and effective fashion while dynamically factoring in developments in the geopolitical context, will be of "make or break" significance during the decade.

Success for this strategy complemented by the continued nurturing of humanistic values and leaders that embrace them would make the 2020s a very good decade for Europe.

References

Moerel, L., Timmers, P. (2021). *Reflections on Digital Sovereignty*, EU CYBER DIRECT, January 2021.

Gallup International. (2021). *Superpowers in the World in Y2030*. February 2021.

Bradford, Anu. (2020). *The Brussels Effect,* Oxford University Press.

DigHumTUWien. (2021). *Preventing Data Colonialism without resorting to protectionism – The European strategy* [Online video]. 23 February 2020. Available at: https://www.youtube.com/watch?v=O7tSwB_3a1w (Accessed: 9 June 2021).

Ministerial Council of the EU. (2020). *Berlin Declaration on Digital Society and Value-based Digital Government at the ministerial meeting during the German Presidency of the Council of the European Union on 8 December 2020.* Available at: https://ec.europa.eu/newsroom/dae/document.cfm?doc_id=75984 (Accessed: 3 May 2021)

Geopolitics and Digital Sovereignty

Ciaran Martin

Abstract The geopolitical dialogue about technology has, for a quarter of a century, essentially revolved around a single technological ecosystem built by the American private sector. An assumption took hold that, over time, clearer "rules of the road" for this digital domain would take hold. But progress toward this has been surprisingly slow; we sometimes refer to "grey zone" activity, because the rules, insofar as they exist, are fuzzy.

In the meantime, the digital climate is changing. China's technological ambitions are not to compete on the American-built, free, open Internet, but to design and build a completely new, more authoritarian system to supplant it. This is forcing a bifurcation of the Internet, and organizations like the European Union and countries across the world have to rethink whether the regulation of American technology is really where the focus should be, rather than working with the USA to contest China's ambitions.

Around a decade ago, it was fashionable in British government circles when discussing the geopolitics of technology to show a picture of Piccadilly Circus in Central London a century earlier. The picture, taken at the dawn of another technological revolution, that of the motorcar, showed various vehicles trapped, pointing in different directions, with no rules to guide who should be going where. The implication was that, over time, humanity learned to develop new "rules of the road" for harnessing technology. Over time, it was asserted, the same would be true of the new digital technologies.

The position is much more complicated now for two reasons. First, setting the rules for the Internet as we understood it then has proved harder and slower than many thought 10 years ago. Second, back then we were talking about a single model of Internet-led technology which was the creation of, by and large, the American private sector. Now there are at least three competing blocs, each seeking supremacy

C. Martin (✉)
Blavatnik School of Government at the University of Oxford, Oxford, UK
e-mail: ciaran.martin@bsg.ox.ac.uk

or at least a competitive position with that model. So before the governance of the American-led model properly settled down, a new geopolitical contest is underway.

Let us first look at the perhaps surprisingly slow progress in the governance of the American-led model. The absence of international rules and standards in many areas of digital life remains a concern for many of America's Western allies. Tax remains problematic: in the summer of 2020, talks between the USA and the European Union on the taxation of digital services broke down; in 2021, the new Biden administration was still challenging the attempts of the UK, now outside the European Union, to introduce a digital services tax which would apply mostly to American companies. The UK, the other non-US Five Eyes, and the EU institutions remain at loggerheads with Silicon Valley over the security and law enforcement implications of end-to-end encryption, with increasingly vocal frustration at the powerlessness of Western governments to reverse the move toward its ubiquity. There have been some movements toward mutually recognized standards in digital trade and data protection. But overall it is hard to claim that the governance of the "free" Internet pioneered on the US West Coast in the late 1990s has progressed much.

Moreover, there are few if any common understandings, let alone rules, about acceptable and unacceptable conduct on that free and open Internet. The first phase of online geopolitical competition has been played on Western or, more precisely, American terms: the USA has most of the infrastructure, the companies, the influence in standards bodies, and much else. So hostile forces, like Russia, when seeking to undermine the USA, are doing so within the digital environment created by the Americans, rather than competing with it. They have exploited its ambiguities and vulnerabilities in a series of what have become known as "grey zone" operations. The zone is "grey" precisely because norms have not been established.

True, the United Nations Open-Ended Working Group on cyber norms unexpectedly reached a unanimous consensus in its third and final round in early 2021. But it is too early to tell whether this compromise will have any lasting impact in terms of the quest for rules of the road in cyberspace. In the meantime, the characterization of a supply chain intrusion for espionage purposes (the so-called Holiday Bear campaign carried out against the SolarWinds company and others) as an act of war by senior members of the US Congress—the sort of activity routinely carried out by Western intelligence services for information gathering purposes—demonstrates the chasm in understanding when it comes to norms. There remains no Western consensus on acceptable activity in cyberspace: for example, Microsoft's quest for a Digital "Geneva Convention" has never attracted serious support from its own government in Washington or any of the USA's most important allies. Western governments, particularly the Five Eyes, show no particular appetite for them.

So the rules of the road remain largely absent. But in the meantime, the geopolitics have evolved significantly. A decade ago, however, one almost universal assumption was that the geopolitics of the new technological age would remain a contest on America's terms. Silicon Valley had no strategic competitor. Moreover, the apparent success of the open and free model was seen as a grave threat to authoritarian regimes seeking to challenge the USA, so shaping the global rules would extend liberal, democratic values.

China has comprehensively overturned in the last decade. Public confidence in the West in its own digital model has wobbled as citizens have increasingly fretted about a range of online harms like online abuse of children, disinformation, monopolistic practices, and cyberattacks. In contrast, and crucially, the first credible challenger model to US-led technology has emerged out of Beijing and the Chinese Communist Party.

The rapid development of Chinese technology has exploded two of the great founding myths of the digital age: that there is no way for states to assert control over the Internet and that the Internet is global and cannot be regionalized. China has long passed the stage where new digital services were a threat to the regime;[1] it has now co-opted new technology into capabilities which further its authoritarian control. And in doing so, it has demonstrated, for example, with its so-called Great Firewall, that the Internet can have borders after all. As Western confidence in America's technology started to wobble, China's self-confidence in its own development, set out with brutal clarity in its 2015 plan known as Made in China 2025 with designs to attain supremacy in many of the key technologies of the future, was published for all to see.

It is now increasingly evident that the USA and China are engaged in an epoch-defining struggle for technological supremacy with huge implications for the geopolitics of the rest of the century. Decoupling of supply chains, already underway with the USA's largely successful campaign to bar Huawei from next-generation telecommunications networks, has now been accelerated by the pandemic. New battlegrounds are emerging: the standards bodies that set technical rules for data transit and infrastructure design and the control of rare-earth metals and other raw materials for hardware among them.

The clash of these two tech titans has brought into sharp focus the relatively new concept of digital sovereignty across various parts of the globe. It can be argued that until recently only the USA was digitally sovereign: in that it could rely pretty much entirely on itself to sustain its digital ecosystem (though there are gaps; the USA's lack of indigenous hardware providers for telecoms infrastructure was exposed during the 5G controversies). But China is becoming digitally sovereign too, not least because the imposition of US sanctions aimed at slowing its technological development has incentivized it never to be reliant on supplies form outside China again.[2]

What of the rest of the world? The most intriguing potential third power bloc is the European Union. The EU is the technological era's great paradox. The USA and China do not really talk about digital sovereignty, but are, by and large, digitally sovereign. The EU talks a lot—and increasingly—about digital sovereignty, but it is

---

[1] For an account of how the Chinese Communist Party wrought control over online communications in China, see *Consent of the Networked: The Worldwide Struggle for Internet Freedom* (Rebecca MacKinnon, 2013).

[2] For a superb account of this trend, see Dan Wang, *Annual Letter, 2019 https://danwang.co/2019-letter/*

very far from digital sovereignty. Two telecommunications equipment giants aside, the European continent of half a billion of the world's wealthiest Internet users has precious little home-grown technological capability. Insofar as it is a tech super-power, it is only a regulatory one. There is obviously an attempt to ground this regulatory posture in digital humanism, but without the industrial capability, this already complicated task is even harder.

Brussels policy is currently facing in two different directions. In July 2020, the German government, in its official program for its presidency of the European Council, announced its intention "to establish digital sovereignty as a leitmotiv of European digital policy" (Pohle and Thie 2020). In its cyber security strategy of the end of that year, the Commission set out, for the first time, some serious ideas on how the development of European technology might take root. But in the same month, the Commission published an overture to the incoming Biden administration proposing economic cooperation on technology, including the establishment of a Transatlantic Trade and Technology Council. This seemed to be a recognition that digital sovereignty in Europe could not be achieved quickly (if it can be achieved at all). Therefore, given the need to align with one of the two genuine technology superpowers, the Americans were the only option, particularly after the departure of President Trump who famously cared little for the USA's alliances in Europe.

Three other more disparate groups of countries will be important in this great geopolitical contest. One, likely to be of great interest to the Biden administration not least because of Europe's ambivalence toward US technology and its current administrative paralysis over coronavirus vaccines, are a set of rich, hi-tech democ-racies across the Five Eyes partnership and Asia (Japan, Singapore, South Korea). None of these have any serious aspirations for digital sovereignty, though Japan and Korea have some serious techno-industrial clout. But they will be keen to align with Washington to counter China's technological ambitions. The challenge is that such an effort does not work like a security alliance: aligning commercial strategies is harder than forming a military pact.

Then there is a group of authoritarian countries who dislike the US technological model every bit as much as China does, but have no capabilities of their own. Russia and Iran are examples of this. They may pursue a version of digital sovereignty which is, in effect, Chinese-style control over the use of the technology without the increasingly Chinese control over the ownership of it. Russia has held some at least partially successful experiments in "disconnecting" itself from the Internet, though a recent test seems to have backfired and hit the Russian government's own infra-structure. In time, such countries may become enthusiastic champions of China's digital model (Russia is already showing signs it may be heading in this direction).

Then, finally, there is the rest of the world: mostly middle- and lower-income countries with no serious expectation of digital sovereignty. Many of these countries increasingly fear being caught between the USA and China and having to choose. For this section of the globe's population which is where much digital growth is expected as gaps in digital inclusion close, the term "digital sovereignty" rings hollow amidst the struggle of the two technological heavyweights. As Deborah M. Lehr of the Paulson Institute put it in 2019, "if a new economic Iron Curtain is to fall, it will be in areas like the Middle East and Africa" (Lehr 2019). Such countries will worry less about the West's ongoing struggles to write rules of the road for America's Internet and more about the implications of this splintering of the technological globe.

References

Pohle, Julia and Thie, Thorsten (2020). "Internet Policy Review" *Digital Sovereignty* 4(9).
Lehr, Deborah M. (2019). *How the US-China Tech War Will Impact The Developing World.* Published in *The Diplomat,* 23 February 2019.

Cultural Influences on Artificial Intelligence: Along the New Silk Road

Lynda Hardman

Abstract Applications of AI, in particular data-driven decision-making, concern citizens, governments, and corporations. China was one of the first countries to have identified AI as a key technology in which to invest heavily and develop a national strategy. This in turn has led to many other countries and the European Union (EU) to develop their own strategies. The societal investments and applications of AI are so far-reaching that looking only at the resulting technological innovations is insufficient. Instead, we need to be aware of the societal implications of AI applications—of which there are many—as well as the geopolitical role of business and academic players.

1 Artificial Intelligence in China

While 20 years ago China was still learning from the international AI community, investments and policies have led to the current situation where China is rapidly overtaking the USA and EU in expertise in AI research, education, and innovation. China has ambitions to become a world leader in AI by 2030 (CISTP 2018), elevating the phrase "made in China" to a data-driven, hi-tech ecosystem for manufacturing goods and technology. China developed an AI strategy in the "China AI Development Report 2018" (CISTP 2018) in conjunction with partners representing both academic and commercial interests in the country, overseen by the

The opinions in this chapter are the author's own and do not necessarily reflect those of her employers or the organizations she represents. This chapter is based on the Amsterdam Data Science blog item "AI Research with China: to Collaborate or not to Collaborate – is that the Question?" and Hardman (2020). https://amsterdamdatascience.nl/ai-research-with-china-to-collaborate-or-not-to-collaborate-is-that-the-question/

L. Hardman (✉)
CWI, Amsterdam, The Netherlands

University of Utrecht, Amsterdam, The Netherlands
e-mail: Lynda.Hardman@cwi.nl

H. Werthner et al. (eds.), *Perspectives on Digital Humanism*,
https://doi.org/10.1007/978-3-030-86144-5_31

233

China Institute for Science and Technology Policy (CISTP) at Tsinghua University. Among the policy goals stated in the report are:

- To increase public awareness
- To promote the development of AI industry (to retail, agriculture, logistics, finance, and reshaping production, e.g.)
- To act as a reference for policy makers

  Societal goals for the use of AI are:

- Helping with an aging population
- Supporting sustainable development
- Helping the country transform economically—toward China as hi-tech developer and supplier, rather than consumer

The first two of the societal goals are shared by Europe and the USA, leading to benefits in collaboration. The transformation of China to a hi-tech supplier is more likely to lead to competition, a valid endeavor in its own right, but results in global competition for talent.

The report is realistic about the Chinese context, stating "Even recognized domestic AI giants such as Baidu, Alibaba and Tencent (BAT) don't have an impressive performance in AI talent, papers and patents, while their U.S. competitors like IBM, Microsoft and Google lead AI companies worldwide in all indicators" (CISTP 2018, p. 6).

The executive summary concludes with "Currently, China's AI policy has emphasized on promoting AI technological development and industrial applications and hasn't given due attention to such issues as ethics and security regulation" (CISTP 2018, p. 7).

China takes a long-term view, and this can be seen in its investments in AI research and innovation, and particularly its tech talent. Huge efforts have been made to attract successful Chinese AI researchers back to their home country to continue their internationally competitive research and to educate new generations of talent. China's presence in the international AI research community is growing, as demonstrated by the increasing percentage of papers in the top international AI conferences that are co-authored by Chinese colleagues, working from China or from abroad (Elsevier 2018).

The CISTP report also observes that the priorities of the USA are economic growth, technological development, and national security (CISTP 2018, p. 5), whereas the concerns of Europe are the ethical risks caused by AI on issues such as security, privacy, and human dignity (CISTP 2018, p. 5). These different regional policies seem aligned with underlying cultural differences among the regions.

2   Artificial Intelligence in Europe

The EU started developing national and European strategies around 2018, for example, establishing the European-wide High-Level Expert Group on Artificial Intelligence,[1] which has produced Ethics Guidelines for Trustworthy AI and corresponding Policy and Investment Recommendations toward sustainability, growth, competitiveness, and inclusion (Craglia et al. 2018), and later publishing a Coordinated Plan on Artificial Intelligence.[2] This drive for AI investment is probably also fueled by the huge investments in China, creating a "fear of losing out" if Europe is not able to remain competitive. The European investment drive is not solely from an economic perspective, as illustrated by the different aspects of the report, such as sustainability and inclusion.

Around the same time that the High-Level Expert Group was developing its report, many computer science academics and professionals were themselves concerned by the growing impact of AI technology and potential unintended negative implications. Informatics Europe and the ACM Europe Council published the joint report "When Computers Decide: European Recommendations on Machine-Learned Automated Decision Making"[3] in February 2018. The report states the utility and dangers of decisions made by automated processes and provides recommendations for policy leaders.

3   Cultural Differences in Applying Artificial Intelligence Technology

While better, and more, researchers across the globe is generally good news for academic research, in AI we need to remain cautious. China's enormous investments in AI have led to domination in a narrow set of sub-fields around machine learning, with an emphasis on computer vision and language recognition. This domination could be perceived as cause for concern from an international standpoint. For example, computer vision techniques can be developed for facial recognition to track the movements of citizens. Different cultures perceive the benefits and dangers of these applications differently. Using these same techniques for other applications, such as distinguishing cancerous from benign cells, is, however, widely perceived as good.

The international community has closely aligned objectives in application areas of AI such as climate change, transport, energy transition, and the health and well-being of their aging citizens.

---

[1] https://ec.europa.eu/digital-single-market/en/high-level-expert-group-artificial-intelligence

[2] https://ec.europa.eu/knowledge4policy/publication/coordinated-plan-artificial-intelligence-com2018-795-final_en

[3] https://www.informatics-europe.org/component/phocadownload/category/10-reports.html?download=74:automated-decision-making-report

This brings us to the difficult political and scientific choices that need to be made as to when and how to collaborate with China and when to politely decline. Do AI researchers need to completely halt all collaboration with Chinese academics and companies? Cessation of collaboration would be counter to the established international research culture of openness and dialogue.

European AI researchers are unlikely to want to work with Chinese colleagues on topics that may aid the Chinese state in actions that do not conform to European civil rights and values. That there is a cultural difference in the desirability ascribed to the trade-offs between privacy and security for those living in China and the West is hard to understand in and of itself and even harder when researchers are not familiar with the Chinese culture. AI researchers are not the most knowledgeable of global cultural differences, nor do many European researchers spend extended periods of time in China to learn first-hand.

Given the relatively small amount of research funding in European countries and from the European Union, the welcome addition of funds from abroad would seem like a golden opportunity. But things are not always as easy as they may appear. Firstly—do European AI researchers want to work with Chinese colleagues? Secondly—do European academic institutions want to be funded by Chinese companies?

It is currently more common for European researchers to collaborate with large US-based corporations. They fund research collaborations and attract high-profile staff to work with them. At the same time, they have created the data economy that led to the passing of EU law to give European citizens at least some control of the data that they (often unknowingly) hand over to these corporations. There is, to my knowledge, little discussion in my academic field as to whether we should think carefully about collaborations with these US-based companies.

4 Artificial Intelligence Talent: Mobility and Global Competition

We need highly educated AI practitioners to develop the wealth of useful/valuable applications across the globe. There is, however, an international shortage of AI talent, including machine learning and data analytics talent. All parts of the globe are looking to educate their own talent and to attract excellent talent from abroad. This requires investment in a number of ways, such as increasing the number of qualified academic staff and increasing the efficiency of teaching. Working for industry brings its own financial rewards. Tenured academic positions are attractive because of the opportunity to carry out research. This requires extended research funding, requiring either a larger slice of the available national (or, in Europe, European) funding or attracting more funding from companies. Companies across the globe are acutely aware of their shortage of talent and are willing to invest in academic research funding to maintain an active base of researchers and teachers in academia. Excellent researchers attract excellent young talent, and, given the English-language foundation of the AI and the computer science fields, the same talent can be attracted to China, Europe, or the USA.

There are, however, differences in the willingness of students to leave their continent to seek their fortune. It is likely that Chinese AI and computer science students will study for some time abroad before returning to China where research resources are currently (anno 2021) plentiful. On the other hand, the attraction of the European work/life balance may play a factor.

European students are much less familiar with Chinese cultures and language than Chinese students are with the English language and American and European cultures. This creates a larger barrier to move to a region where the currently perceived academic benefits are few. This may change as awareness of the technological speed of change and available research resources in China increases, creating a stronger pull for both European and American students.

While China is attractive to young hi-tech talent, working weeks are long, giving little opportunity to spend time with family and friends. In Europe, it is not just one's standard of living that is important but also one's quality of life. Development of AI technologies provides hope that both can be achieved through more efficient use of the limited resources available.

Huge efforts have been made to attract successful Chinese AI researchers back to their home country to continue their internationally competitive research and to educate new generations of "home-grown" talent. Both generations of researchers bring with them the competitive, individualistic risk-driven culture learned abroad. Just as in winning in top sport—be it gymnastics, football, or table tennis—two things are essential: individuals with intrinsic potential and motivation and an environment that polishes and hones the required internationally competitive skills. A characteristic of the successful Western research culture is questioning received wisdom, which goes against the grain of Chinese and many other Asian cultures where authority is highly respected.

5   Global Collaboration on Artificial Intelligence Research and Innovation

So what are my recommendations in this complex and sometimes contradictory collaboration puzzle?

China is a world leader in AI research, technology, and innovation. As its investment into this field continues to grow, this will only become more pertinent. We therefore cannot ignore the relevance of China in AI research and development, but we can be considered in our approach to collaborations and make informed decisions on a case-by-case basis. Bekkers et al. (2019) provide concrete guidelines on how to approach collaborations.

Questions of international AI collaborations and their effect on the global population are large and daunting. Learning more about your Chinese collaborators and colleagues is a much easier and definitely more pleasant task. Read the Lee (2018) book, which gives insights into taking the Silicon Valley start-up culture and transferring it to China while at the same time metamorphosing it to the rules of a new "Wild East." Learn Chinese and visit your colleagues in China.

AI research and innovation is taking place along the New Silk Road. European researchers are already accustomed to global collaborations across different cultures. One of the characteristics of international research culture is the independent exchange of critical feedback, at the same time remaining aware of the implications of the research outcomes. Chinese researchers are developing their own research and innovation strategies, making strategic investments in academic education and research enabling them to become an influential partner on the global stage in research as well as innovation. From both European and Chinese perspectives, AI researchers need to develop a better understanding of the cultures in which we operate. Learning from each other's cultural perspectives is something that our AI systems are not yet able to do for us.

References

Bekkers F., Oosterveld W. and Verhagen P. (2019) Checklist for Collaboration with Chinese Universities and Other Research Institutions. The Hague Centre for Strategic Studies, January. https://hcss.nl/report/checklist-for-collaboration-with-chinese-universities-and-other-research-institutions/

CISTP (2018) China AI Development Report 2018. China Institute for Science and Technology Policy (CISTP) at Tsinghua University. http://www.sppm.tsinghua.edu.cn/eWebEditor/UploadFile/China_AI_development_report_2018.pdf

Craglia M. et al., 2018. 'Artificial Intelligence: A European Perspective', Publications Office of the European Union, March. https://ec.europa.eu/jrc/en/publication/artificial-intelligence-european-perspective

Elsevier (2018) Artificial Intelligence: How knowledge is created, transferred, and used. Trends in China, Europe, and the United States. https://www.elsevier.com/research-intelligence/resource-library/ai-report

Hardman L. (2020) Artificial Intelligence along the New Silk Road: Competition or Collaboration? Chapter in: (Van der Wende, 2020). https://ir.cwi.nl/pub/29940

Lee K.-F. (2018) AI Superpowers: China, Silicon Valley, and the New World Order. Houghton Mifflin Co., USA. https://www.aisuperpowers.com

Van der Wende M.C., Kirby W.C., Liu N.C. and Marginson S. (eds.) (2020) China and Europe on the New Silk Road: Connecting Universities across Eurasia. Oxford University Press. https://global.oup.com/academic/product/china-and-europe-on-the-new-silk-road-9780198853022

Geopolitics, Digital Sovereignty...
What's in a Word?

Hannes Werthner

Abstract An overlay of digital networks and services often operated by global players encircles and "shrinks" the planet. At the same time, the geopolitical dynamics have entered a cycle of feud for leadership between trade blocs who compete for economic and industrial leadership but also on ethics, values, and political outlook. In this context, governments and lawmakers are struggling to combine the need for global cooperation in digital matters with the imperative to protect their jurisdiction from undue influence and provide economic agents with the means to compete on a global scale. The concept of "digital sovereignty" was carved to address this. Words matter a lot especially when they are meant to translate political goals. We argue that "digital sovereignty" lacks meaning and teeth, while the concept of "strategic autonomy" is more operative, contains in itself the elements of strategic planning, and should lead EU to aim at genuine "digital non-alignment."

1 The Context

1.1 The Paradox

Nations (and groupings thereof), industry, and international organizations have spent megabillions in the last decades to connect the globe. Sub-marine cables, satellite communications, fixed and mobile networks, the Internet, enabled global connectivity and access in the most remote places.

To enable this, the geo-political dynamics shaped a world of multi-lateral collaboration, with the WTO[1] as the arbiter of a frictionless global trade. The global village

---

[1] The World Trade Organization is an intergovernmental organization which regulates international trade. The WTO officially commenced on 01/01/1995 under the Marrakesh Agreement, signed by 123 nations in 1994, replacing the General Agreement on Tariffs and Trade (GATT), which commenced in 1948 (www.wto.org).

H. Werthner (✉)
Vienna University of Technology, Vienna, Austria
e-mail: hannes.werthner@tuwien.ac.at

© The Author(s) 2022
H. Werthner et al. (eds.), *Perspectives on Digital Humanism*,
https://doi.org/10.1007/978-3-030-86144-5_32

was not a rose garden indeed, but multi-lateral regulatory cooperation would fix it, and global trade would make the world a safer and wealthier place for all. So was the narrative, for example, when China became a member of the WTO in 2001.

Fast forward to 2021, "second life" has become our life, digital connectivity has infused through the economy and society worldwide, M-Turks from Nigeria or India work in real time for Silicon Valley corporations, and geography is ended (so is privacy, but this is a different subject). And yet, as technology, economics, and politics have shaped this global cyber-reality, globalization is challenged by the (re)formation of more or less antagonistic trade blocs. After three decades of moving toward a single global market governed by the rules of the WTO, the international order has undergone a fundamental change, and an open, unified, global market may indeed become a thing of the past (Fischer 2019).

1.2   It's More Than the Economy, You Know

With different narratives, each regional trading bloc is developing its own roadmap to achieve global digital success and indeed global supremacy. Be it President Biden signing an executive order strengthening the "Buy American" provisions, China asserting its primacy in digital matters and global trade, India promoting a techno-nationalistic agenda, or Russia developing its offensive cyber capabilities, it is as if global trade in the twenty-first century was bound to be a discordant zero sum game.

What's more, this global competition is not only industrial, technological, or economic but also about visions, values, and methods. Whether trade irritants can dissolve in good intentions remains to be seen.[2]

In a conference of the Centre for Economic and Policy Research in February 2021, Dr. Christian Bluth[3] argued that the most important challenge for EU today was the increasingly charged geopolitics of trade. He argued that trade policy is increasingly used for projecting power rather than generating prosperity and several countries are "weaponizing" the trade dependence that others have on them.

2   Europe, How Many Divisions?

Be it for tea in China, spices from the Malabar Coast, gold in South America, coffee or rubber in Africa, and oil in the Middle East, Europe did not have a problem with global supply chains or national sovereignty when it was dominant in worldwide trade and industry. It even got support from moral or political authorities (Treaty of

---

[2] Some call for "differentiated digital sovereignty."

[3] https://www.bertelsmann-stiftung.de/en/about-us/who-we-are/contact/profile/cid/christian-bluth-1

Tordesillas in 1494,[4] Berlin Conference, 1885[5]) and carved the political concepts to lean on, e.g., Westphalian sovereignty. Today's situation is indeed slightly different.

2.1 A Pacific Centered *Digital* World Map

Once an economic giant (but a political dwarf), Europe's ambitions for the "digital decade" are caught in between a duopoly with the USA and China dominating the global digital economy. *The Old Continent appears to have already lost the artificial intelligence battle—to name but one. We need to wake up to the fact that we are falling behind in 5G development, and its application in service and industrial verticals, and so running the risk of becoming a minor player in the global contest.*[6]

2.2 In the Platform Economy, Nobody Can Hear EU Scream

GAFAMs (a short for Google, Amazon, Facebook, Apple, and Microsoft) and BATs (Baidu, Alibaba, Tencent) are not only global platforms with revenues much larger than many countries' GDP. They also integrate vertically and horizontally, absorbing potential competition, shaping the whole economy including for strategic sectors and the provision of public services. Incidentally, they also alter the fundamentals of the labor market. This market is not EU's strong suit as shown in Fig. 1.

2.3 Digital Sovereignty, a New Concept to Operate and Compete in This Context

Originating in the cybersecurity community, the concept of "digital sovereignty" gained ground in lawmaking circles with the increasing number of attacks on critical infrastructure (power, communications, water, etc.), the global connectivity, and the sky-rocketing number of IoT devices with its related security and privacy issues. The complex relationship with the previous US administration also contributed to propagate the idea that EU had to rely on its own capacity to defend itself in the hyperconnected world.

---

[4] An agreement between Spain and Portugal aimed at settling conflicts over lands newly discovered by Christopher Columbus and other late fifteenth-century voyagers.

[5] Conference at which the major European powers negotiated and formalized claims to territory in Africa

[6] IDATE, Digiworld 2020 https://en.idate.org/the-digiworld-yearbook-2020-is-available/

Top 100 worldwide platforms

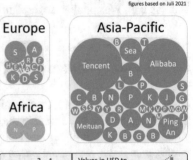

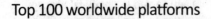

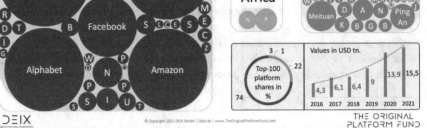

Fig. 1 The imbalance of platform economy (H. Schmidt, TU Darmstadt, https:// TheOriginalPlatformFund.de/, 2021)

Devising new fundaments to build EU policy in this challenging context, the European Commission—and some Member States—developed the concept of "digital sovereignty," defined by Commissioner Breton[7] as:

- Sovereignty on data, especially industrial data (a sovereign cloud)
- High-performance computing and microchips
- Connectivity (5G, optical fiber, and low-orbit satellites)

With its Communication on 2030 Digital Compass issued in March 2021[8], the European Commission defined "the European way for the Digital Decade" to translate EU's digital sovereignty objectives into specific targets. Additional policy instruments such as the Communications on Industrial Strategy[9] specify the path toward EU leadership in digital.

---

[7]EU Commissioner for industrial policy, internal market, digital technology, defense, and space
[8]https://bit.ly/3l0MBH4
[9]https://bit.ly/3aWrv94

3 Words Matter: Especially When They Are Meant to Be Performative

3.1 Digital and Sovereignty, How Does This Add Up?

The digitalization of the world adds a meta layer on top of the political authority. To set the rules of the game in its own jurisdiction, EU policy makers devised a series of regulatory measures: General Data Protection Regulation,[10] Cybersecurity Act, Directive on Network and Information Security,[11] Digital Services Act, and Digital Market Act.[12]

The legal framework is set—and this is not trivial, but does it suffice to build EU's capacity to be sovereign in the digital competition? Many EU companies that play in the global league are not particularly fervent of the concept of sovereignty as most of their operations and revenue are overseas. How about the indecision with GAIA-X[13] constituency or ARM (a British leading chip maker) acquired by its US rival NVIDIA ("a disaster for Cambridge, for the UK and for Europe" H. Hauser BBC Radio 4 in September 2020) and envisaging to subsidize a US company to build EU chip industry capacity?

3.2 Political and Legal Considerations

Defined by F.H. Hinsley (1986, p. 1), sovereignty is "the idea that there is a final and absolute political authority in the political community [. . .] and no final and absolute authority exists elsewhere." This implies, on the one hand, that no political authority can be half sovereign and, on the other hand, that the entity from which sovereignty emanates should be monolithic or at least sufficiently integrated to project "final and absolute political authority." Both characteristics are in contradiction with the way the EU is constructed and the breakdown of jurisdiction and competence between EU and Member States.

Does the exclusive competences of the EU[14] grant EU lawmakers the means to walk the talk? What means sovereignty without jurisdiction? How sovereign when, for example, 15 EU countries representing over 50% of the entire EU membership

---

[10] https://bit.ly/3nKbzvW

[11] For the Cybersecurity Act and NIS Directive, see https://bit.ly/3gYlE6M.

[12] For the Digital Services Act and the Digital Market Act, see https://bit.ly/2QIfHR2.

[13] GAIA-X is a project for the development of a competitive, secure, and trustworthy federation of data infrastructure and service providers for Europe, supported by representatives of business, science, and administration from Germany and France, together with other European partners (https://bit.ly/3nD241q).

[14] Customs union, competition rules for the internal market, monetary policy for the euro area, common fisheries policy, common commercial policy, and conclusion of international agreements

sign with China's Belt and Road Initiative or European automotive brands ink deals with the GAFAs for data analytics, machine learning, and artificial intelligence?

This might partly explain the bids from the European Commission to seize activities in areas the Treaties allot in principle to Member States, such as radio spectrum allocation, health, or e-identity. Maybe "life on life's terms" will change this breakdown of competence; for the time being, the EC's bids have not been welcomed with open arms in EU capitals.

4 Where Next?

4.1 Many Assets to Mobilize

EU may not be a leading player in several areas of the digital economy, e.g., platforms. Yet it has a series of assets to build on, such as:

- The largest GPD in the world and a market of 500 million
- Leadership in several domains (e.g., aeronautics, cryptography, banking, automotive retail)
- A very dynamic SME scene
- R&D and intellectual capital
- High-end connectivity and networks (transport, energy, telecoms)
- Cultural diversity and a fundamental rights charter

Europe is a deep wellspring of talent, with a tremendous capacity to rebound, and a rare power of innovation: (. . .). Europe is also synonymous with actions and projects driven by exacting values and a commitment to positive and progressive construction.[15]

4.2 Strategic Autonomy

We argue that leadership in digital does not mean leading in all segments of the tech industry, but rather the capacity to digitalize industry and services in a safe, secure, and trustworthy way. With this come the questions of how to combine those assets, which battles to choose, which allogenous bricks can be part of the plan, who to partner with on what terms, etc.

In other words, select strategic sectors, and within those, make one's own rules and own plans on its own terms. Strategic autonomy, literally, as in *auto-nomos*. This implies acting in several directions and selecting what to do and what NOT to do—which is often the forgotten part.

---

[15] IDATE, Digiworld 2020, op. cit.

This is not new. To a large extent, the EC (and several capitals) are headed in this direction with the recent Regulations mentioned above, the increased powers granted to ENISA[16] in cybersecurity, or the rules to participate in EU-funded research projects.

Contrarily to the concept of digital sovereignty, the concept of strategic autonomy does not hint at any notion of protectionism, rather the idea that "you're most welcome to operate in my jurisdiction as long as you play by the rules I set." It is also much more operational as it almost self-contains the notion of a dynamic planning.

In a webinar of a Brussels think tank in February 2021, Anthony Gardner, ex-US Ambassador to the EU, said in a very poetic manner that "Digital sovereignty as sometimes heard in EU circles is chasing moonbeams." Strategic autonomy on the contrary is very down to earth and operational.

4.3 Aim for the Moon

In the debate on digital sovereignty—or strategic autonomy as we prefer to call it, China is the geopolitical elephant in the room. In order to fight on the geopolitical scene in a heavyweight (industrial) category, Europe often advocates to "partner with think-alike," those who share EU values on freedom of speech, democracy, and human rights. This makes sense and is possibly much encouraged by the new US administration.

The other elephant in the tech and economic room are the GAFAMs. And as described earlier, all EU trade partners ponder the ambition to be global leaders in the digital world and act accordingly.

In this context, we argue that Europe could define a "third way," a sort of "digital non-aligned" doctrine which would also give it total leeway to make its strategic choices and rules:

- Define areas for collaboration and no-go areas (e.g., in research programs, in FDI).
- Keep friends close and enemies even closer (that's what diplomacy is for).
- Respect historical allies while controlling lobbying and entry strategies.

The continent has all it takes to embody this third way and build its own path in the otherwise digital bipolar world shaping before us. This might seem utopian but so was the European construction at its beginnings when the founding fathers launched the process in a devastated continent, which interestingly started with industrial cooperation on coal and steel, as essential to the economy then than digital is today.

---

[16]European Union Agency for Cybersecurity, https://www.enisa.europa.eu/

Acknowledgment This chapter was written after a long discussion and exchange with a third person who prefers to remain anonymous.

References

Hinsley, F.H. (1986). *Sovereignty*. 2nd edition. Cambridge: Cambridge University Press.
Fischer, J. (2019). 'The End of the World As We Know It' *Projekt Syndicate*. 3 June [online]. Available at: https://www.project-syndicate.org/commentary/us-china-break-europe-by-joschka-fischer-2019-06 (Accessed: 15 June 2021)

Part IX
Systems and Society

Work Without Jobs

Daniel K. Samaan

Abstract Technology has always had an impact on the world of work. This chapter compares the transformation of our societies during the Industrial Revolution with potential transformations that digitalization may bring about today. If digitalization is truly disruptive, more may be at stake than job losses in some sectors and job gains in others. Identifying several key features of digitalization, this chapter sketches a future of work in which not jobs but work itself stands in the center of economic activity. Such a development could open a pathway to more humanistic, more democratic, and more sustainable societies but would require rethinking entirely how we organize and reward work on a societal level.

Around 200 years ago, many societies in Europe and North America fundamentally altered the way in which work was organized and remunerated. Facilitated by technological advances like the steam and combustion engines, as well as expedited by regulatory changes, mass production and standardization of goods became the prevailing modes of production. This newly emerging factory system also entailed changes in the work organization: it has been characterized by a high physical concentration of labor in production facilities and a hitherto unseen division of labor, orchestrated by hierarchical organizations. Both changes, mechanization and standardization of production processes and the corresponding new work organization, have led to unprecedented productivity gains to which we owe much of today's living standards. In his famous example of the pin factory, Adam Smith has illustrated the magnitude of such productivity increases more than two centuries ago, whereby output per worker could be increased to 4800 pins from less than 20 (Smith 1776).

We all know today that historians would later refer to this decades-long period of continuous and fundamental changes to the world of work as the Industrial Revolution (IR). Closely tied to this revolution is what Frithjof Bergmann (2019) calls the

D. K. Samaan (✉)
International Labour Organization (ILO), Geneva, Switzerland
e-mail: samaan@ilo.org

© The Author(s) 2022
H. Werthner et al. (eds.), *Perspectives on Digital Humanism*,
https://doi.org/10.1007/978-3-030-86144-5_33

"job system": We bundle the vast majority of our work activities ("tasks") into "jobs." We call standardized descriptions of such jobs "occupations." These jobs are then bought and sold on the (labor) market for a supposedly fair share of society's overall output (wage). Hence, the functioning of the industrial society is centered, not about work that we do for us, but about obtaining and performing jobs for others. The question I want to pursue in this chapter is whether in a "digital society" this interdependence will be any different.

The importance of the "job system" for our societies can hardly be underestimated. It is at the center of how we act and how we conduct our lives: We educate ourselves, predominantly, in order to "learn an occupation" and to "get a job." We want to spend our lives being "employed" and not "unemployed." Being "unemployed" and without a "real job" are social stigmata and lead to loss of income and social standing. Political competition in every Western democracy is critically concerned about creating new jobs or proposing suitable conditions for companies to crank out more jobs. We are prepared to accept all kinds of unpleasant trade-offs, like destroying our environment or heating up the climate, if only job creation is not harmed. Because without jobs, we have no work, no income, no taxes, no public services, no social security systems, no more support for democratic votes, and finally no more society, as we currently know it.

This way of thinking has not changed much since the IR. In 2021, we do reflect about the future of work, but our imagination of the future is restricted and dominated by the "job system" and by all the institutions and terminology that we created around it: "the labor market," "re-skilling," "unemployment," "part-time work," "contingent work," etc. This list could be easily expanded and filled with the respective literature on the future of work. In other words, with some exceptions (e.g., Precht 2018), most of the discussion on the future of work sees the job system as a given centerpiece of our societies.

The job system has not always existed. In pre-industrialization times, working from home or in small community shops on one's own terms, self-controlled and owning the means of production, was the norm. Several factors drove us into the creation of the "job system" at the time.

How does digitalization figure in this debate? It has awakened old fears among workers, politicians, corporate leaders, middle managers, and others. Specifically, they worry that this most recent wave of digitalization will lead to unprecedented automation and hence a massive loss of jobs (Frey and Osborne 2017). And as we have seen above, once the jobs are gone, the downward spiral (no work, no income, etc.) is triggered. So, this fear is justified.

Yet, I would like to look at this question from a slightly different angle in this chapter: A society might run out of jobs, but it can never run out of work. The real question that we face today is therefore whether or not digitalization and its powerful offspring, big data and artificial intelligence (AI),[1] are going to eradicate the "job system" and, if so, how we can live without it.

There are three reasons why digitalization, understood as a technology, has the potential to destroy the job system. Firstly, artificial intelligence is a general-purpose technology (Brynjolfsson and McAfee 2014). It is not an invention, like the radio or many others, which have a confined impact on certain economic sectors and societal domains, like the radio has had on mass media, the printing press, and perhaps the military sector. AI, and digitalization more broadly, is more comparable to electrification. We can find applications and devices in virtually all economic sectors for consumers and producers, workers, management, governments, and many other actors alike. This qualification as a general-purpose technology is a major ingredient for a revolutionary change. The economic system is shocked from many different contact points at the same time.

Secondly, big data provides economic actors[2] with information on the "states of the world" and facilitates decentralized decision-making and decentralized action. Most of economic activity on the societal level (often also on the individual level) is about making decisions under uncertainty to allocate resources efficiently: A priori, we do not know who needs which goods and services under what conditions at a certain time. Neither do we know who can supply which resources under which conditions and who has which capabilities. We do not know and cannot directly observe in which "state" the world is. This was the price we had to pay for the high division of labor. Traditionally, this lack of information and this problem of coordination have been solved by adding middlemen and by mass-producing a standardized good or service for the average consumer. Those "middlemen" can be persons inside an organization, like middle managers, who pass information from the top management to the workers and make sure the orders are carried out. The middlemen can also be outside an organization (say a firm) and facilitate contact to the right customers or carry out marketing surveys. The information flow and the feedback are typically coordinated through specified channels that follow hierarchical structures. These are remains of the factory system. Digitalization makes much of this framework unnecessary. Production plants and workers do not need to be concentrated, neither spatially nor in time. Output does not have to be standardized but can be customized for a specific individual. We can think about the industrial economic world as a picture of islands of producers, customers, workers, and managers, whereby the middlemen are connecting the islands. The whole picture ("state of the world") is not fully visible. Now big data is rapidly filling the empty spaces with many small dots and establishing direct connections among them.

---

[1] Digitalization encompasses more than the massive amounts of digital data and processing of information through AI systems, but I will have mainly these two aspects in mind when I refer to digitalization in this chapter.

[2] In fact, not all economic actors have the same access to information, and the accumulation of digital data over the last years has already led to a power shift across enterprises.

Thirdly, digitalization brings about an enormous potential to automate tasks. Such automation can but does not have to result in the automation of occupations, as Frey and Osborne (2017) claim. Brynjolfsson et al. (2018) develop a methodology to analyze about 2000 so-called detailed work activities (DWA). These DWA build the backbone of the 1000 occupations that currently exist in the United States. The authors calculate a metric for each DWA that measures to what extent the activity is suitable to machine learning (SML), and since an occupation is just the weighted average of a subset of the DWA, one can also come up with an SML score for each occupation. They find that many US occupations are suitable to machine learning. Unfortunately, such a comprehensive study does not exist for work activities and occupations in countries other than the United States. Furthermore, the analysis is based on existing tasks and on work processes as they exist today. These processes and hence (human) tasks were of course designed in a manner to reflect the capabilities of the currently existing technology (i.e., different technology = different work processes and different tasks). Therefore, one needs to be careful before interpreting the study's results as a prediction for job losses in the US economy or to generalize it to other countries. Yet, the authors demonstrate convincingly how digitalization can lead to a rapid decomposition of a large number of occupations and of existing jobs into underlying tasks. For companies, workers, and governments, this development can lead to a fast destruction of jobs, the emergence of new jobs, and/or a reorganization of work and "re-definition" of jobs.

Taken these three above features of digitalization together, I think, it is very well possible that we will see more than a mere reshuffling of a few lost jobs in some sectors and emerging new jobs other ones, as we have seen in the past. There is potential for—literally—a "digital revolution" in the world of work rather than a slight and continuous adjustment of the status quo. This could become problematic if our societies attempt to cope with this challenge by sticking to the paradigm of the "job system." It seems, however, that this is exactly what policymakers and—to a lesser extent—also enterprises and their people are doing. I am not aware of any country that has proposed a truly new vision for the future of work, other than "lifelong learning" (or "re-skilling"), anemic phrases that have been in use for more than 30 years now. Will that be enough to steer changes in the world of work toward more humanistic, democratic, and sustainable societies?

Going back to my initial statement that a society can never run out of work, this may be a reassuring fallback position; it really cannot, at least theoretically. A society can, however, fail in the coordination of its work potential toward the satisfaction of its material and humanistic needs. This is why we have clung to the job system in the first place: It has provided political stability, kept the masses busy, and provided most of us with an abundance of commodities. It has also created a notion of distributional fairness in presenting itself as a meritocracy.

What is wrong then with the job system? Put bluntly: Almost everybody hates it, and we all know that it is at most partially fair. An increasing number of people feel caught in "bullshit jobs" (Graeber 2018) where they perform objectively or subjectively useless tasks, many of which could be done by machines or only exist to justify the person's "job." Very few people in the workforce do work they "really

really want" (Bergmann 2019), even though the desire to work and contribute to the well-being of society in one way or another exists in virtually every human being. Is digitalization not giving us the tools to devise a better, more human coordination and remuneration mechanism? According to some estimates, the percentage of jobs in "transaction industries"[3] has risen from about 15% in 1900 to about 40% in 1970 in the US economy. Unfortunately, I do not have more recent, comparable numbers available, but these estimates were from before the big wave of "servicification" hit our economies. I would therefore estimate that today more than 2/3 of the jobs are "transaction jobs," including a large overlap with what Graeber (2018) calls "bullshit jobs." Karl Marx (1885) referred to labor that is essentially spent to circulate capital and goods as "unproductive labor," compared to "productive labor" that is concerned with the creation of use values for society.[4] If we adopt this notion of productivity, only a small proportion of all performed labor is still productive for society. Interestingly, going back to the more modern works of Frey and Osborne (2017) and Brynjolfsson et al. (2018), we can see that the jobs/tasks with a high risk of automation and high suitability for machine learning are exactly these jobs/tasks in the "transaction industries." Finally, the job system, conjointly with mass production, has a terrible record in terms of resource productivity. It is a waste producer.

A society with more meaningful and satisfying work is possible. Most of us want to spend more time on care work, education and cultural work, and preserving and protecting our environment. The digital revolution is not threatening our work, it is threatening the job system, and this is good news.

References

Bergmann, Frithjof (2019): 'New work, new culture – work we want and a culture that strengthens us', Zero Books, Winchester U.K, Washington U.S.

Brynjolfsson, Erik, and Andrew McAfee (2014): 'The Second Machine Age: Work, Progress, and Prosperity in a Time of Brilliant Technologies'. Reprint. W. W. Norton & Company.

Brynjolfsson, E., T. Mitchell, and D. Rock (2018): 'What Can Machines Learn and What Does It Mean for Occupations and the Economy?' AEA Papers and Proceedings, 108, 43-47.

Frey, Carl Benedikt, and Michael A. Osborne (2017): 'The Future of Employment: How Susceptible Are Jobs to Computerisation?', Technological Forecasting and Social Change 114: 254–80. https://doi.org/10.1016/j.techfore.2016.08.019.

---

[3] Sectors and professions concerned with processing and conveying information (accountants, clerks, lawyers, insurance agents, foremen, guards, etc.).

[4] Marx did not use the terms "productive" and "unproductive labor" consistently throughout all three volumes of capital.

Graeber, David (2018): 'Bullshit Jobs – The rise of pointless work and what we can do about it',
    Penguin Books, UK, USA.
Marx, Karl (1885): 'Capital', Volume 2, Penguin Edition of 1978, 1885.
Precht, Richard David (2018): 'Jäger, Hirten, Kritiker – Eine Utopie für eine digitale Gesellschaft',
    Goldmann Verlag, 1. Auflage, München.
Smith, Adam (1776): 'An Inquiry into the Nature and Causes of the Wealth of Nations', Modern
    Library; 12/26/93 edition (December 26, 1993), New York.

Why Don't You Do Something to Help Me? Digital Humanism: A Call for Cities to Act

Michael Stampfer

Abstract Cities across the globe face the challenge of managing massive digitization processes to meet climate goals and turn urban agglomerations into more livable places. Digital Humanism helps us to see and define how such transformations can be done through empowerment of citizens and administrations, with a strong political agenda calling for inclusion, quality of life, and social goals. Such an approach appears to be much more promising than top-down technological fantasies as often provided by large companies in fields like housing, transport, the use of public space, or healthcare. The title refers to a question put to Stan Laurel by Oliver Hardy in countless movies. Here the latter stands for a city calling industry for help. The delivery as we know can lead straight to disaster, but in real life it is less funny than with the two great comedians.

Cities across the world face a number of pressing *and* long-term challenges, including massive urbanization with growth in size and density as well as the need to de-carbonize the whole urban metabolism. This includes transport, construction, and consumption as well as implementation of climate mitigation strategies. Further, and in many different ways, cities play an important role in key policy areas like social cohesion, education, housing, and health, with the aim to provide for an affordable and high quality of life. Finally, as most real politics is local, cities are pivotal for further developing democracy, sourcing the creative potential, fostering innovation, and increasing the political participation of their inhabitants.

In the last decades, digitalization has been entering through all kinds of doors, with great promises and massive power to transform traditional forms of evidence-gathering, business models, governance structures, communication patterns, and decision-making. Ubiquitous optical and sensor systems and broadband networks provide for massive and reliable data which can be analyzed with powerful methods ranging from machine learning to complex systems analysis. In fields like health or

M. Stampfer (✉)
Vienna Science and Technology Fund (WWTF), Vienna, Austria
e-mail: michael.stampfer@wwtf.at

© The Author(s) 2022
H. Werthner et al. (eds.), *Perspectives on Digital Humanism*,
https://doi.org/10.1007/978-3-030-86144-5_34

public transport, the power to collect, own, combine, and interpret data has become as important as the ownership of operating theaters or subway lines. Cities across Europe therefore speed up with digital policies and actions. Speed however is a relative term as many cities face huge obstacles with endless layers of bureaucracy, laws cementing the status quo, conflicting interests, and limited budgets.

Now we can close our eyes for a moment and think of a stressed-out, impatient Oliver Hardy turning to Stan Laurel, inconspicuously standing close by: "Why don't you do something to help me?" This is how many cities have acted: Most conveniently, industry also happens to wait already on the doorstep with beautifully rendered turnkey solutions or mobile apps to solve most wicked societal problems. Unfortunately, for miracles like *thumb-as-lighter*, *finger-wiggling*, and *kneesy-earsy-nosey*, the transfer of skills to the unprepared mind has its consequences: We let our imagination still flow for a moment, to uphill piano transports, escalating cream pie fights, or the sweet dynamism of destroyed kitchen porcelain. At the end, we listen to a sobbing Stan Laurel pleading innocent and to the famous Oliver Hardy line: "That's another nice mess you've gotten me into." However, contrary to Stan, industry often does not end on the losers' side.

"Smart City" has become a key term for urban policies dealing with data-driven, often large-scale solutions to better manage urban agglomerations. For cities to enter the next steps of digitalization means facing a number of challenges; the following five points can be seen as examples:

A first one might be just termed "turnkey"—like supplying whole neighborhoods with readymade supply for de-centralized energy production and storage with smart grids and meters. This is a very good idea (seriously, and at the same time we hear Oliver Hardys' voice again), a big trend, and it helps transform the energy system, but it is also extremely tricky and in need of long and patient co-development between private and public actors.

A second challenge is "you may keep the hardware," as the tram network or bus fleet stays with the public utilities, but the data solutions managing the user interface are being serviced by private providers. That one might not be such a good idea, as data today defines strategies and directs revenue streams. Therefore, cities try to establish their own data management structures.

A third one can be termed "improved, electrified, without driver, no change of mind required" which is a nice combination of current user habits and future industry profits. Take as an example the effort to help cities get rid of the car pandemic . . . by providing technically enhanced cars: As without drivers they for sure will after the trips miraculously disappear somewhere. A better idea might be to establish demand-based co-creation processes with citizens on how to redefine and regain urban space and broader mobility concepts.

A fourth challenge is the "law-overriding platform economy," successfully ignoring taxation, employment laws, or sector regulations, just because they can. When cities or regions hit back with their still powerful old-world instruments like taxi license ordnances, regulations for touristic accommodation, or labor standards, they might fend off the tech giants for a while, albeit at the price of stifling innovation and paying rents to incumbents.

A fifth example is "give us your city and make *us* happy," with again tech giants collecting and sucking off all kinds of data without leaving sustainable profits for the city and its inhabitants. Many examples start with low capabilities of city actors to frame, organize, manage, and capitalize digital platforms and services, therefore handing over to industrial actors. They then provide street view, health diagnosis, or pan-optical surveillance, without telling what they do with the data collected and how they re-use and sell them.

Such examples hopefully show how important it is for cities to develop and implement an active political approach toward data and data policies. This starts with focusing on data protection and privacy issues as well as with building up in-house competencies, extending to the creation and nurturing of local networks and knowledge hubs, as well as cooperation with academia and civil society. Strategies are important, and more so are scores of individual projects, ranging from supported small citizens' initiatives to large-scale change efforts in health, transport, energy, or participation.

The bottom-up approach is specifically important: Cities have to become active and experiment, for three reasons: First, in many fields connected with large-scale data collection, the platform economies, and their digitally driven influencing strategies on our future behavior, we see national, European, or global regulation still at an infant stage, with large companies successfully battling legislative efforts. Second, for changes in the way cities work and resources are being used, the ideas and needs of citizens are often the best way to reclaim public space, data sovereignty, and carbon neutrality. Data sovereignty is of specific importance here, as city administrations are also hungry for data and should not misuse them. Third, living labs and trial and error are often much better-suited for lasting innovation than the top-down, turnkey solutions. As we of course also need general laws and regulations, such an approach can serve as a valuable learning space.

Digital Humanism plays a central role in shaping what cities do and how they do it, as a state of mind and as a guiding principle. City politics and administration do not have to re-invent the wheel as there are many successful examples in history how to create public goods in areas like housing or public transport. The "state of mind" issue appears to be of specific importance as the question is: Who shall prosper? How to include all—or at least most—inhabitants in decision-making and have them share the benefits? What is good life in cities in key elements like social interaction, health, resource consumption, or space to live?

These are not really new questions; take the example of a city like Vienna in pre-digital times: Powerful public infrastructures, a huge communal housing program, top-class health for all, and strong social networks have been the political priorities for more than 100 years. As one consequence, Vienna in the last decade could successfully take the next step and become one of the leading Smart Cities across the globe, combining goals and measures for social inclusion, for innovation, and for reduction in resource consumption and carbon emissions.

The next, ongoing step is to frame digitalization politically *and* in all policies to respond to the *Digital Humanism number one question*: How can we preserve and constructively transform the best of our current civilization into the digital world?

We have to build laws, norms, structures, and knowledge bases to allow our key institutions to thrive also in the future: This is the representative democracy, the welfare state, the rule of law, and the social market economy. All of them can unfold even better in a strongly digitized world. However, leaving this transformation without strong political action, we invite big market actors to suck off our data, to not pay taxes, or to manipulate elections and our individual actions. Without such frameworks, we write an invitation letter to illiberal democracy and to the surveillance state, as unfortunately the libertarian equation of greater freedom through unregulated digital interactions has proved to be a two-edged sword at best.

Digital Humanism therefore means active politics and regulatory frameworks in addition to ethics, knowledge bases, and infrastructures. Cities can do a lot in all these respects:

- Active politics is about setting an agenda, finding majorities, and making issues visible through priority setting and sometimes loud, bold statements. Mayors and cities are more powerful than one might be aware of. They can mobilize an electorate and are in charge for big decisions in city development. A number of cities like Barcelona or Amsterdam have already implemented policies that certain data sets collected by big companies have to be made available to citizens, small companies, and initiatives. Cities can play an important role in guaranteeing individual and collective data sovereignty and help make data a resource for social and economic purposes on a local level. Cities can form coalitions and pressure groups, as with the Digital Services Act on the European level.

- As in former industrial revolutions, regulatory frameworks in many ways still wait to be developed. In the meantime, policy makers also on the city level have to respond by shifting the objective: As already stated, such regulations might not always be rooted in the digitization as such. Instead, we find them in good old labor laws, transport licenses, or accommodation rules. As a consequence, Uber-style transport might come to an abrupt halt; and Airbnb hosts find unexpected obstacles. Such measures help stop the erosion of justified standards in labor relations or fair competition. They might at the same time also stifle competition, feed lazy incumbents, and prevent customer value from materializing. Therefore, we need new forms of regulation, and cities can be strong actors between a Charybdis called good old times and the Scylla of unregulated global platforms devouring scores of workers and businesses.

- On the basis of strong political and regulatory will, ethics issues come in through many doors. A first door is education, with new curricula to help the gap between the "lots of ethics but no idea about technology" approach in the more humanistic colleges and the "give us points to connect and gratifications and progress at all price" ideology of technical education. Such an integration should start early on, and regional authorities can play a role as they often run the more basic education. A next door is the purchasing power of cities asking for high ethical standards in all kinds of digital goods and services. A third one is supporting grassroot and civil society actors concerned with privacy, data sovereignty, or creative work; cities can subsidize and promote movements, festivals, neighborhood initiatives, and a score of other activities. One final point here goes in the same direction, supporting a critical discourse, where the ethics part is about taking the side of the less fortunate part of the population by countering a narrative like: "we all have to be techno-optimists" by stating that without caring mainly the strong and powerful will collect the benefits when this narrative materializes.

- Knowledge bases—besides schools—include strong universities and research providers. Many cities employ active policies to support their local research base. Cities like Berlin have created cross-disciplinary centers, while in Vienna we find a research funding program called Digital Humanism to link Computer Sciences with Social Sciences and the Humanities. Here the idea is to collaboratively create theories, methods, approaches, and practices as well as to find a common understanding of new technological and social phenomena. Such initiatives shall help to wake up the soft sciences for the challenge of the digital revolution, while it shall provide the engineers with a framework of *how* a good, inclusive society shall further unfold with the help of their models and artifacts. One guidepost for such activities is the *Vienna Manifesto for Digital Humanism* that has been co-developed by local and international researchers of various backgrounds with support from Vienna policy makers.

- Infrastructures is another broad topic where cities can play an important role: First by helping to provide top-class broadband networks, second by strongly supporting their own departments and utilities as well as across private industries to come forward with up-to-date solutions, both technically and non-technically. Third, and perhaps most important, is the ability of the public sector to effectively deal with their data. Currently public actors collect loads of data but often without proper policies and practices how to best store, validate, connect, and share them. Within the framework of the recent European data protection regulations, there are many ways to better steer policy, deliver results, and allow research to access data. Unfortunately, many public actors including cities currently face a dilemma: As they cannot always effectively transform traditionally high standards of service into the digital sphere, they have to give carte blanche to all kinds of companies including the global platform firms by letting them collect, analyze, and capitalize the data. Examples from the health and public transport sector show that such an approach can be dangerous: Public actors should at least have the competence to govern public domain data, being able to decide what shall remain in the public domain and what can be handed over to the private sector.

As we see, Digital Humanism is a mindset and a tool for cities, a mindset in emergence and a tool in the making. We all can be part of this process. *Why don't you do something to help us?* for now is a serious question.

Further Reading

Bria, F., https://www.youtube.com/watch?v=9i2dZgbsagY. (Francesca Bria talking about what has been going on in Barcelona and other cities regarding data sovereignty and giving back data value to the people)

City of Vienna, https://smartcity.wien.gv.at/en/approach/framework-strategy/ (The Vienna Smart City strategy and approach can be found here)

Digital Humanism Initiative Vienna, https://dighum.ec.tuwien.ac.at/dighum-manifesto/ (This is the Vienna Manifesto on Digital Humanism)

Hal Roach Productions, https://www.youtube.com/watch?v=DiFEFL6ThRI (Stan Laurel showing Oliver Hardy *kneesy-earsy-nosey*, from the movie Fra Diavolo)

Vienna Science and Technology Fund (WWTF), https://www.wwtf.at/digital_humanism/index.php?lang=EN (WWTF funding activities)

Weizenbaum Institute, https://www.weizenbaum-institut.de/en/ (Weizenbaum Institute in Berlin presents itself)

Weizenbaum Institute, https://www.youtube.com/watch?v=B9_EeHjNcVE (video with presentations of the research groups)

Zuboff, S., https://www.youtube.com/watch?v=fJ0josfRzp4 (A long lecture by Shoshana Zuboff on Surveillance Capitalism)

Ethics or Quality of Life?

Hubert Österle

Abstract Governmental and non-governmental organizations around the world are trying to shape socio-technical development, especially the use of information technology, for the benefit of people. They are developing ethical guidelines for the creation and evaluation of digital services. The discipline of Life Engineering must combine the knowledge of several disciplines, such as psychology, machine learning, economics, and ethics, so that technology serves people, i.e., contributes to well-being. Therefore, a solid understanding of quality of life should be the starting point, explaining the patterns of human behavior and their impact on well-being. Digital services of all kinds provide increasingly detailed digital twins and give us the opportunity to operationalize ethical principles.

For decades, machine intelligence has changed companies and the economy. Now it is affecting our lives in significantly more direct ways and generating hopes and fears. Ethical initiatives such as Digital Humanism want to align machine intelligence with the quality of life (happiness and unhappiness) of all humans.

1 Abundance and Fear Determine the Discussion

For highly developed societies at least, technology and capitalism have brought enormous material prosperity and satisfied needs such as food, security, and health, i.e., the needs of self-preservation and preservation of the species.

But the affluent society can do more than satisfy basic needs (yellow background in Fig. 1). The needs of selection (light-blue background) come to the fore (Österle 2020, p. 68–80) and drive human beings onto a treadmill in which, consciously or unconsciously, everyone is constantly working on their status, whether through

H. Österle (✉)
University of St. Gallen, St. Gallen, Switzerland
e-mail: hubert.oesterle@unisg.ch

© The Author(s) 2022
H. Werthner et al. (eds.), *Perspectives on Digital Humanism*,
https://doi.org/10.1007/978-3-030-86144-5_35

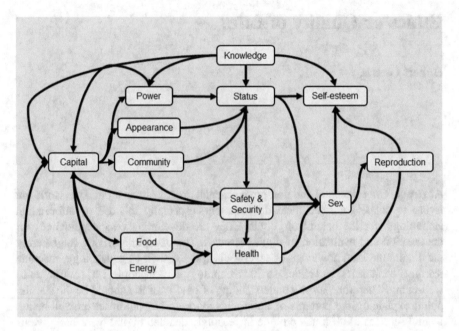

Fig. 1 Network of needs

clothing, offices in a club, knowledge, skills in music, youthful fitness, or simply
through capital. An almost explosive growth in the literature on happiness research
and ethics as well as an accompanying offer of lifestyle services such as happiness
training, yoga, and wellness is aimed at helping us to gain as many positive feelings
as possible from the satisfaction of all needs and to avoid negative feelings.

At the same time, there is a growing fear of what is to come. Dystopias such as
surveillance capitalism (Zuboff 2019), the totalitarian surveillance state, the loss of
humanity and traditional values, or the excessive demands placed on the individual
distract from the urgent task of shaping the coming change.

2 Development Requires Ethical Guidelines

Phrases such as "for the benefit of humanity" have become a common element of
corporate mission statements. But who actually believes in such grandiose statements? What has ethics, especially business ethics, as formulated by Max Weber
100 years ago (Weber 1915), actually achieved? It is certainly helpful to ask what
kind of interests guide ethics.

3 Companies and Business Leaders Want to Satisfy Their Stakeholders

At the American Business Round Table, nearly 200 CEOs of leading US companies signed a "fundamental commitment to all of our stakeholders."[1] Many media articles have described it as an attempt to sugarcoat the social ills of digitalization through simple declarations of intent. Interestingly, the statement of these business representatives does not even mention the much more concrete international standard ISO 26000 on Corporate Social Responsibility (Schmiedeknecht and Wieland 2015), which was adopted 10 years ago. Digitalization requires many corporate leaders to demonstrate, among other things, the responsible handling of personal data. Individual management consultants have reacted to this with offers for data ethics, aimed primarily at maintaining company ratings.

4 Investors Are Looking for Returns Through Sustainability

Investors seek additional financial performance through investments that meet environmental and social criteria as well as the requirements of good governance (ESG—environment, social, and governance). They want to identify the opportunities and risks of their investments at an early stage based on these criteria and thus increase the profitability of their investments. Rating agencies like MSCI[2] and inrate[3] evaluate listed companies according to ESG criteria for investors. In accordance with the recommendations of the OECD,[4] politicians use the weight of the financial markets to achieve sustainable development.

5 Do-Gooders Misuse the Ethics Discussion

Avoiding the dangers of digitalization and seizing the opportunities for the benefit of human beings is a task for all citizens. Everyone must consider how they use digital services and what they expect from companies and politicians, for example, what personal data they give to Facebook, and where politicians should protect them from abuse. The danger arises when the discussion is dominated by do-gooders, who often

---

[1] https://www.businessroundtable.org/business-roundtable-redefines-the-purpose-of-a-corporation-to-promote-an-economy-that-serves-all-americans

[2] https://www.msci.com/documents/10199/123a2b2b-1395-4aa2-a121-ea14de6d708a

[3] https://www.inrate.com/index.cfm

[4] https://www.oecd.org/finance/Investment-Governance-Integration-ESG-Factors.pdf

argue purely emotionally, usually represent a very narrow partial view, and use vocal debate to compensate for their lack of knowledge and thus influence politics. Typical "enemies" are the greed of shareholders, the totalitarian manipulation in China, the taxation of foreign corporations, and the "zombification" of mobile phone users. Do-gooders altruistically stand up for the good of the community but demand sacrifices mostly from others. In many cases, their commitment is a search for recognition for their efforts and a striving for self-esteem, which is often described as a "meaningful life" or similar phrases.

6   Politics Follows the Need for Ethical Rules

Politicians need votes or the trust of their constituents. So they pick up on the popular mood and translate it into pithy catchphrases. A good example is the European Union's announcement of the digital future of Europe[5] with populist values such as fairness, competitiveness, openness, democracy, and sustainability. In addition to emphasizing fashionable topics such as artificial intelligence, the chapter focuses on the regulation of digitalization, while it hardly presents any concepts on how Europe should keep pace with the USA and China and therefore actively contribute toward shaping digital services. The focus is on restricting entrepreneurial activity, not on exploiting potentials such as the Internet of Things (5G, sensor, and actuator technology). The addressed citizens do not know these technologies or know them too little, and they have neither the time nor the motivation and the prerequisites to understand the technologies and their consequences. It is therefore much easier to evoke the previously mentioned "bogeymen" than to arouse enthusiasm for misunderstood technologies.

This is also confirmed by the discussion on the use of mobile phone users' location data to curb the spread of COVID-19. The data that has long been used, for example, for planning public transport, is virtually negligible compared to the use of data voluntarily submitted to Google, Apple, or Facebook. Even classic personal data such as the traffic offenders' register in Flensburg, credit scorings, and customer data in the retail sector allow for far more dangerous misuse. Ethical values cultivated by do-gooders and attention-grabbing media hamper any serious discussion on how the rapidly growing collections of personal and factual data could help to make human coexistence healthier, less conflictual, and more enjoyable,[6] rather than concentrating on tightening criminal law.

---

[5] https://ec.europa.eu/commission/presscorner/detail/de/ip_20_273
[6] https://www.lifeengineering.ch/post/social-scoring-the-future-of-economy-and-society

7 Ethics Wants Quality of Life for All

Ethics is looking for rules that should bring the highest possible quality of life for everyone. If we accept that digitalization cannot be stopped and that it will bring about massive socio-cultural change, we need mechanisms, now more than ever, to guide this change for the benefit of humankind. But do ethics and the underlying interests provide the tools? Two essential prerequisites are missing: First, ethics does not determine what quality of life actually constitutes. Second, there is a lack of procedures for objectively measuring quality of life.

A discipline called Life Engineering should start right there. It should develop a robust quality of life model based on the findings of psychology, neuroscience, consumer research, and other disciplines and validate this model using the increasingly detailed and automatically collected personal and factual data. The network of needs can be a starting point if each of the needs, like health, is broken down into its components, such as age, pain, weight, strength, and sleep quality, and the causal relationships are statistically recorded.

Once the factors of quality of life are better understood, it will be possible to better assess the opportunities and risks of digital services. The sensors of a smartwatch can measure possible influencing factors on health so that individualized correlations between physical activity and sleep behavior or heart rhythm disturbances can be recognized and the wearers of smartwatches can thus increase their health and well-being by taking simple measures. Such concrete, statistically sound evaluations of digital services currently remain the exception. However, a quality of life model, even in such a rudimentary form as the network of needs outlined above, provides at least a framework for discussion in order to evaluate technical developments in terms of arguments, as shown by the example of Instagram.

Ethics is based on values such as dignity, respect, trust, friendship, responsibility, transparency, and freedom. However, such values are only relevant to people if they meet their needs and thus trigger positive or negative feelings. What does the ethical value trust mean for needs like security, power, or energy?

It very quickly becomes clear how far away we are from a quality of life model that combines behavior, perceptions, needs, feelings, and knowledge. However, looking at the tasks of ethics, it is hardly justifiable not to at least try what is feasible. Right now, we are leaving this development to the Internet giants, who, like Google, for instance, with its knowledge graph, try to better understand and model these connections, while these companies and their management are being measured by their economic success, not by human quality of life. It is therefore almost inevitable that they will have to persuade customers to make the decisions that generate the most revenue.

Never before in the history of humankind have we had such comprehensive and automatically recorded datasets that allow statements about behavior and quality of life. The Internet and sensors are documenting our lives more and more seamlessly, as Melanie Swan discovered as early as 2012 under the banner of the "quantified self" (Swan 2012, p. 217–253). The instruments of machine learning and modeling in neural networks offer us the chance to recognize quality of life patterns and to make them effective in digital assistants of all kinds, from shopping to nutrition, for the benefit of human beings. Never before has such intensive support been provided for people by machines in all areas of life through digital services. Never before has it been possible to give people such well-founded and well-targeted help and advice, to guide them in a recognizable but subtle way. The thought of this frightens the pessimists and excites joyful expectation among the utopians.

With the methods of data analytics, health insurance companies evaluate the personal and factual data of their policyholders in order to better calculate the individual risks. They adjust the individual premiums in line with the individual risks and ultimately reduce claim costs for the same income. For some policyholders, this leads to savings, but for those who are disadvantaged in terms of health and therefore financially less well-off at the same time in most cases, it means higher payments. The redistribution of risk in the sense of solidarity is lost.

If an insurance company succeeds in better understanding the influences on health and—what is even more difficult—in guiding the insured to health-promoting behavior through digital services, then this machine intelligence helps both the insured and the insurers.

8   Ethics Needs Life Engineering

Development cannot be stopped, but the direction can be influenced. We need a discipline called Life Engineering that translates the humanities concepts of traditional ethics and philosophy into design-oriented proposals, i.e., that pragmatically shapes technical, economic, and social development.

Only those who drive and lead development can influence it. The aversion to technology, which can be felt in many ethical discussions, has exactly the opposite effect to what it aims to achieve. It is therefore extremely welcome that scientists from engineering and technical sciences, from social science and humanities, for example, in the DIGHUM initiative of Vienna University of Technology (TU Wien) or the Ethically Aligned Design[7] initiative of the IEEE (Institute of Electrical and Electronics Engineers), are coming together to formulate rules for machine intelligence. Even without an elaborated quality of life model, it is possible to avoid at least some clearly unwanted characteristics of digital services. This calls among other things for rules stipulating that people can access and check the data stored about them and approve its use or that a machine decision must be justified. However, these rules come up against the limitation of human cognitive abilities, i.e., whether a layman can even understand these connections within a reasonable time.

---

[7] https://standards.ieee.org/content/ieee-standards/en/industry-connections/ec/autonomous-sys tems.html

Apart from these obvious rules, which do not have to be derived from scientific studies, it would be helpful if ethics could be based on an operational quality of life model. It is positive that version 2 of the IEEE guidelines on Ethically Aligned Design, unlike the first version, attempts to do just that. It is based on approaches and metrics for well-being. Its recommendations on the different aspects of ethics for machine intelligence ultimately provide a comprehensive agenda for Life Engineering.

In order to ever be able to meet such requirements, a Life Engineering discipline needs the following, in addition to financial resources:

- Access to the digital personal and factual data
- Exchange of knowledge about behavior patterns and their effects on quality of life
- Ability to influence the development of digital services
- Political incentives for positive developments and prohibitions of negative developments

Life Engineering offers the chance to transfer ethics from the stage of a religion to a stage of science, just as the Enlightenment did in the eighteenth century. This has brought about a human development that probably only few people today would like to reverse.

References

Österle, H. (2020). *Life Engineering – Machine Intelligence and Quality of Life*. https://doi.org/10. 1007/978-3-030-31482-8, p. 68-80.

Swan, M. (2012). 'Sensor Mania! The Internet of Things, Wearable Computing, Objective Metrics, and the Quantified Self 2.0'. *Journal of Sensor and Actuator Networks*, 1(3), 217–253. https:// doi.org/10.3390/jsan1030217

Schmiedeknecht, M. H. & Wieland, J. (2015). *ISO 26000, 7 Grundsätze, 6 Kernthemen. In Corporate Social Responsibility. Verantwortungsvolle Unternehmensführung in Theorie und Praxis.* Berlin, Heidelberg: Springer Gabler.

Weber, M. (1915) *Die Wirtschaftsethik der Weltreligionen.* Jazzybee Verlag.

Zuboff, S. (2019). *The Age of Surveillance Capitalism.* New York: PublicAffairs.

Responsible Technology Design: Conversations for Success

Susan J. Winter and Brian S. Butler

Abstract Digital humanism calls for new technologies that enhance human dignity and autonomy by educating, controlling, or otherwise holding developers responsible. However, this approach to responsible technology design paradoxically depends on the premise that technology is a path to overcoming human limitations while assuming that developers are themselves capable of super-human feats of prognostication. Recognizing developers as subject to human limitations themselves means that responsible technology design cannot be merely a matter of expecting developers to create technology that leads to certain desirable outcomes. Rather, responsible design involves expecting the technologies to be designed in ways that provide for active, meaningful, ongoing conversations between the developer and the technology, between the user and the technology, and between the user and the developer—and expecting that designers and users will commit to engaging in those conversations.

Digital humanism calls for new technologies that enhance human dignity and autonomy by infusing ethics into the design process and into norms and standards. These calls are even echoed by politicians in the international arena (Johnson 2019):

> the mission ... must be to ensure that emerging technologies are designed from the outset for freedom, openness and pluralism, with the right safeguards in place to protect our peoples. ... we need to agree on a common set of global principles to shape the norms and standards that will guide the development of emerging technology (Boris Johnson, Address to the UN, 2019)

Although we should strive to achieve this vision of responsible technology design, we should also realize that we cannot meet it. Technologies are created by people. Both technologies and people are limited. Because of these limitations, the best we can hope for is that new technologies allow for ongoing dialogue that recognizes and responds to the need for human dignity and autonomy.

S. J. Winter (✉) · B. S. Butler
University of Maryland, College Park, MD, USA
e-mail: sjwinter@umd.edu; bsbutler@umd.edu

© The Author(s) 2022
H. Werthner et al. (eds.), *Perspectives on Digital Humanism*,
https://doi.org/10.1007/978-3-030-86144-5_36

While there are many nuances and variations, one of the simplest ways to understand the foundational premise of digital humanism is to consider a digital humanist in contrast with a digital technologist. Both want to create a better world and improve the human condition. Both see technology as a way of overcoming human limitations and making things better.

Where they differ is that digital technologists see the creation of technologies that eliminate the need for human involvement through automation as the primary path to sustained, substantial improvement in the human condition. Self-driving cars seek to remove the need for a human driver. AI-based facial recognition seeks to remove the human from the recognition process. In contrast, digital humanists pursue change by encouraging development of technologies that place humans at the center, empowering them to enact their own well-being. Wikis enable humans to create collections of useful information for efficient learning. Social platforms allow people with similar goals to create mutually supportive communities of practice.

However, while they differ significantly with respect to the role of humans in the application of technology to improve the human condition, one thing that many digital technologists and humanists share is an assumption about the relationship between developers and the technologies they create. Whether it is implicit or explicit, it is often assumed that developers create technologies which in turn shape the actions and choices of the users (Gibson 1977). The actions of developers lead to the different technologies existing (or not), having particular features (or not), and creating affordances which enable (or prevent) users from taking particular actions or making particular choices (e.g., Anderson and Robey 2017).

It is this assumed power of the developer to shape the technology and subsequent user behavior that is the basis of efforts to bring about responsible technology design by educating, controlling, or otherwise holding developers responsible. Efforts to infuse ethics into Computer Science curricula reflect this assumption, for example, Embedding Ethics[TM] @ Harvard which:

> embeds philosophers directly into computer science courses to teach students how to think through the ethical and social implications of their work. (https://embeddedethics.seas. harvard.edu/)

The premise is that if we nurture developers' appreciation of ethics, the self-driving cars that they create will be safe and efficient, expand mobility to under-served communities, and create a better, more equitable world. If we sensitize developers to the need for privacy, they will build privacy protections into AI for facial recognition that will enhance personal and societal security without intrusive over-surveillance. Educated, thoughtful, ethically aware developers will design wikis that reject unverified and inaccurate information. Developers who are aware of the ethical and social implications of their work will encourage regulation that will result in platforms that support pro-social communities of practice and block those that are pursuing goals of violence and hate.

Somewhat paradoxically, this approach to responsible technology design is based on the premise that technology is a path to overcoming human limitations while also assuming that developers are themselves capable of practically super-human feats of prognostication and influence. Ethically trained, sensitized, well-regulated developers will still be surprised by how, when, and why their "responsibly designed" self-driving cars, facial recognition software, wikis, and social media platforms will be deployed. Contexts are infinitely variable, and users are infinitely "creative." Assuming that a developer (or a user for that matter) will "get it right" drastically overestimates humans' ability to imagine, anticipate, and influence the functioning of even the most basic socio-technical systems.

Responsible technology design cannot be merely a matter of expecting developers to create technology that leads to certain desirable outcomes. To posit this definition of responsible design necessarily requires a capability that is beyond the reach of any human designer and will lead to expectations about developer responsibilities and obligations that are at best unreasonable and at worst dangerously misguided.

Rather, responsible design involves expecting the technologies to be designed in ways that provide for active, meaningful, ongoing conversations between the developer and the technology, between the user and the technology, and between the user and the developer and expecting that designers and users will commit to engaging in those conversations. It is well within our ability to create systems and technologies that provide the affordances for iterative designer-tech, user-tech, and designer-user conversations. Indeed, in the face of the human limitations outlined above, the only forms of responsible technology design that are feasible are those based on repeated iterative, active, adaptive engagement with the technology by developers. Instead of defining success as developers creating a responsible design, we must expect that they engage in the never-ending process of responsible design.

One common approach to enabling these conversations involves developers incorporating (and committing to using) affordances that allow them to collect and attend to data about the technology. Self-driving cars record their state and action. AI facial recognition technologies track classification choices. Wikis support tracking of content changes and user actions. Social platforms track blocked content. This is seen by some as the minimum required for responsible technology design. Yet ultimately this approach still assumes that developers will have the somewhat super-human ability to use the data to review, monitor, track, and interrogate the performance and use of the technology while also having the control needed to not misuse the data.

As the quote from Boris Johnson above indicates, concerns about the outcomes of new technologies are not limited to the simple performance of the technological components of systems. Responsible design implies responsible engagement with the larger socio-technical system and the processes by which meaning, purpose, and values emerge. Developers will have to build in (and commit to using) affordances that allow them to collect and attend to data about not just the technological component, but the larger socio-technical system in which it is embedded. Records of actions by self-driving cars are important, but equally so are the choices of the driver. Responsible design of AI and facial recognition requires attention to issues of accuracy, but also to issues of appropriateness in their application. Responsibly designed wikis must filter misinformation and disinformation, but also must choose

S. J. Winter and B. S. Butler

how to balance the desires of readers with the inclusion of silenced voices and peoples. Social platforms must track the volume of posts and types of content, but must also continually consider trade-offs between the economic goals of the providers and the civic goals of the larger society. Engaging in dialog with a system requires that developers engage with these issues as well while balancing the needs for privacy and security.

These conversations can occur at multiple levels and in diverse forms. An agile or co-design approach creates a direct dialog between users and designers that is much richer than is possible with the waterfall method of development. Regulation also puts users, developers, and technologies in conversation with one another. Users can also use the marketplace to express their preferences. Of course, responsible technology design cannot mean that a developer is responsible for all of the outcomes of the technology. Rather, we argue that they are responsible for creating and engaging in systems that support ongoing dialog, engagement, and adaptation between developers, technological elements, and other stakeholders.

By their very nature, digital technologists necessarily set themselves up with a fundamentally harder problem with respect to enabling responsible design. By setting their sights on eliminating meaningful involvement of humans in systems through automation, they necessarily make the dialog between developers and those systems more difficult to support and achieve. Building in traceability, detailed logs, exception reports, and an extensive investigative operation to review and respond to this data in a timely fashion become essential. At best, building this capability for a more responsible system requires substantial additional cost and effort. At worst, it requires developers to incorporate features and functions which are counter to the goals of automation, setting responsible design up in opposition to what a digital technologist considers to be an effective design process.

In contrast, a digital humanist who seeks to improve the human condition by empowering people is already predisposed to enabling dialog between the human and technical element of a socio-technical system because that dialog is integral to their approach. Incorporating additional features and functions that enable developers to participate in this dialog is therefore a more straightforward proposition and less likely to be seen as counter to the goals of the design process. Digital humanism can help us recognize the limitations of humans and the role that technology can play in empowering humans to overcome those limitations. This is a significant contribution that digital humanism can make. Recognizing the limitations of developers and users and adopting models of responsible design and use that accommodate those limitations by putting these communities in continual conversation could be an even more powerful contribution of digital humanism.

Whether it is self-driving cars enabled by the internet of things, artificial intelligence for facial recognition, self-monitoring wikis, online community platforms, or some other application of emerging technology, it is not the consequences of designers who failed to anticipate the impact of their creations that we should fear. This we must expect even with extensive training in ethical decision-making. There is no other outcome that is possible. Instead, it is the designer who has the hubris to believe that they can fully anticipate the outcomes of their creations—and as a result fails to allow for and participate in the conversations that are needed to adaptively engage the technology and its implications that are the irresponsible parties who should be the object of concern.

References

Anderson, C. and Robey, D. (2017). Affordance potency: Explaining the actualization of technology affordances, Information and Organization, 27(2), 100-115, 2017.

Embedding Ethics[TM] @ Harvard, https://embeddedethics.seas.harvard.edu/, retrieved April 15, 2021.

Gibson, J. L. (1977). A Theory of Affordances. In R. Shaw and J. Bransford (Eds.) Perceiving, Acting and Knowing: Toward an Ecological Psychology, Hillsdale, NJ: Lawrence Erlbaum Associates, Inc. pp. 67-82.

Johnson, B. (2019). Prime Minister speech for the UN General Assembly, Sept, 14, https://www.gov.uk/government/speeches/pm-speech-to-the-un-general-assembly-24-september-2019.

Navigating Through Changes of a Digital World

Nathalie Hauk and Manfred Hauswirth

Abstract In this chapter, we address the question of how trust in technological development can be increased. The use of information technologies can potentially enable humanity, social justice, and the democratic process. At the same time, there are concerns that the deployment of certain technologies, e.g., AI technologies, can have unintended consequences or can even be used for malicious purposes. In this chapter, we discuss these conflicting positions.

Information technologies have become an integral part of work, health, entertainment, communication, and education. Yet, the great hope of this technological (r)evolution of opening up a world of possibilities—unlimited access to information, free expression for all, clean energy, sustainability, economic growth, and industrial innovations—has turned into a fear of living under a surveillance state with "transparent" citizens. Increasingly, society is divided into those who consider themselves as progressive, willing to jump on the bandwagon of technological innovation, and those for whom things are moving too fast, who feel powerless in defending their rights and safety. This debate within society has been called the "midlife crisis of the technological revolution" (Ars Electronica 2019), referring ironically to the search for orientation of people in their 40s. Consequently, this is the right moment to ask ourselves the fundamental questions of "why we develop technologies" and "what purpose they serve." Historically, humans have always used tools to overcome their own (physical) limitations to ensure survival. Today, digital technologies carry the potential to not only overcome physical limitations but to promote and enable humanity, social justice, and the democratic process. For that, it is crucial to address

N. Hauk (✉)
Fraunhofer Institute for Open Communication Systems (FOKUS), Berlin, Germany
e-mail: nathalie.hauk@fokus.fraunhofer.de

M. Hauswirth
Technical University of Berlin, Berlin, Germany

Weizenbaum Institute, Berlin, Germany
e-mail: manfred.hauswirth@tu-berlin.de

the issues of morality, ethics, and legality in the development of technologies since the ultimate limit of technologies must be the ethical and moral limits. For more details on the topic, see chapter "Our Digital Mirror" in "Ethics and philosophy of technology."

A prime example of the increasing reliance on technology in the modern society is Artificial Intelligence (AI). AI algorithms and technologies have already found their way into everyday life. Hence, the question of whether AI technologies should be employed no longer arises. The performance of routine tasks, such as using web search engines, opening a smartphone with face ID, or running automatic spell checks when writing an email, relies on AI, often unnoticed by the user. Nevertheless, as with any new technology, the use of AI brings both opportunities and risks. While AI can help with protecting citizens' security, improving health care, promoting safer and cleaner transport systems, and enabling the execution of fundamental rights, there are also justified concerns that AI technologies can have unintended consequences or can even be used for malicious purposes.

Good exemplary demonstrations of these fundamental problems can be found in the area of Machine Learning (ML). ML is used to discover patterns in data, e.g., identifying objects in images for medical diagnoses. The big advantage is clear: An ML algorithm never gets tired and performs the tedious analysis task for enormous numbers of images, at high frequencies and speeds. With the invention of quantum computers, even larger amounts of data could/will be analyzed in real time, tackling problems that are out of reach until now. However, ML algorithms are often "black boxes"—capable of performing a learned or trained behavior without offering insight into how or why a decision is made (for a brief overview on explainable AI, see Xu et al. 2019). For the ML training process, the appropriate selection of training data sets is of crucial importance. The deployment of inappropriate or biased data sets often only becomes apparent after a training process has already been completed, as the following three examples illustrate:

1. Automated decision-making processes are deployed increasingly in recruiting and human resources management. In 2018, Amazon had to cease an AI recruiting tool after discovering that the underlying algorithms of their software discriminated against women. Presumably so because the initial training data set contained more male applicants than female applicants. Hence, the algorithm learned that the best job candidate was more likely to be a male.
2. Tay, a chatbot developed to research conversational understanding, released on Twitter by Microsoft in 2016, started using abusive language after receiving vast quantities of racist and sexist tweets from which Tay learned how to conduct a conversation.
3. An image recognition feature, developed by Google in 2015, miscategorized two black people as gorillas. The fact that the company failed to solve the problem (but rather blocked the categories "gorilla," "chimp," "chimpanzee," and "monkey" in the image recognition entirely) demonstrates the extent to which ML technologies are still maturing.

A product must ensure the same standard of safety and respect for fundamental rights, regardless of whether its underlying decision-making processes are human-based or machine-based. Moreover, we create AI systems that are able to write texts and can communicate in natural language with us. Some of them do this so eloquently that we are no longer able to distinguish whether a real person communicates with us or a system. This fundamental problem has been shown by Joseph Weizenbaum with his simple natural language processing system ELIZA (Weizenbaum 1967). Since then, such systems, e.g., virtual assistants, have evolved significantly, while the problem still remains unsolved. The potential danger is that we do not know whether a given piece of information comes from a human or a machine. Thus, we cannot infer the reliability of a given information, or we may have to re-define the concept of reliability altogether. These issues become even more delicate and pressing when fundamental rights of citizens are directly affected, for example, by AI applications for law enforcement and jurisdiction. Traceability of how an AI-based decision is taken and, therefore, whether the relevant rules are respected is of utter most importance.

1   Trust as a Key Driver

Fears of surveillance and malicious use of technology potentially decelerate or even prevent technological and societal progress. So the fundamental question is:

> How can we increase trust in technological development in order to generate value from the application of technologies?

Trust is a key antecedent of ensuring technology acceptance (Siau and Wang 2018) and, thus, a key requirement for continuing the progress of technological development. This is particularly important, when dealing with technologies that are not directly controlled by humans or if they make autonomous decisions. The importance of trustworthy AI has been identified and emphasized also at the political level, e.g., by the European Commission (European Commission 2020). It is important to distinguish between trustworthiness of a technology, e.g., AI, and trust in technologies. While trustworthy AI is comprised of normative ideas on the qualities and characteristics of a technology (that may or may not depend upon ethical considerations), trust in technologies is based on psychological processes through which trust is developed (Toreini et al. 2020). Yet, the concepts of trustworthiness of AI and trust in AI are intertwined. The concept of trustworthy AI is based on the idea that trust builds the foundation for sustainable technology development and that the full benefits of AI deployment can only be realized if trust can be established. At the same time, addressing the ethical considerations in the process of technology development or deployment influences the formation of trust, such as confidence that systems are designed to be beneficial, safe, and reliable.

Definitions of trust can have different emphases depending on the type of trust relationship, e.g., trust toward individuals, toward organizations, or toward machines

(Bannister and Connolly 2011). The comparability of the concept of trust in an interpersonal relationship and trust in machines is subject to ongoing scientific debate. Either way, the general concept of trust always includes a perception of risk, e.g., any type of negative consequence that might derive from using a technology. Trust can be defined as "the willingness [. . .] to be vulnerable to the actions of another party [. . .] irrespective of the ability to monitor or control that other party" (Mayer et al. 1995, p. 712). The formation of trust in technology specifically depends on the interplay of three characteristics: (1) human characteristics, such as personality and abilities; (2) environmental characteristics, such as morals and values of a given institution or culture; and (3) technology characteristics, such as the performance of the technology, its attributes, and its purpose (Schäfer et al. 2016; Siau and Wang 2018). Generally, the influence of human characteristics and environmental characteristics will be similar regardless of the type of trust relationship. Therefore, we continue with the specificities of technology characteristics in the human-technology trust relationship.

Human interaction with technology is increasingly moving away from the simple use of computers as tools toward building relationships with intelligent, autonomous entities that carry out actions independently (deVisser et al. 2018). As technological devices become ever-more sophisticated and personalized, the way humans bond with technology, e.g., by touching and talking to machines, intensifies as well. People have the tendency to anthropomorphize technology and, in the case of AI, to apply human morals to it (Ryan 2020). However, to apply human moral standards to machines is problematic since not even very complex machines like AI technologies possess consciousness, intentions, or attitudes at the moment (and possibly never will). Nevertheless, research suggests that the formation of trust in technologies depends on the level of perceived "humanness" of a technology—the perception of human-like traits, e.g., voice and animation features (Lankton et al. 2015). Furthermore, people develop trust in technologies in different ways, e.g., along more human-like criteria or more system-like criteria. According to the ABI+ model of trust (Mayer et al. 1995; Dietz and Den Hartog 2006), characteristics that enhance perceived trustworthiness in a person are ability, benevolence, integrity, and predictability. *Ability* refers to the skills and competencies that enable the trustee to have influence or deliver a desired outcome. Corresponding system-like criteria would be technical robustness and safety of a technology. *Benevolence* refers to the belief in the goodwill of the trustee. Applied to technologies, the perceived level of benevolence can be increased by the responsible deployment of a technology and its transparency. *Integrity* refers to a set of principles of the trustee that is perceived as respectable. To increase the level of perceived integrity in technologies, we must strengthen their reliability and accountability. Finally, *predictability* refers to the stability of perceived trustworthiness sustained over time.

For example, people are more likely to trust a new technology when that technology is provided by an institution with a high reputation—representing ability, benevolence, integrity, and predictability—in contrast to an institution without such a reputation (Siau and Wang 2018). Trust in chatbots, for example, depends partially on the perceived security and privacy of the service provider (Følstad et al. 2018).

Furthermore, if technologies are perceived as reliable, transparent, and secure, the trust in a technology increases (Hancock et al. 2020).

2 Conclusions

Society relies increasingly on technologies to stay competitive and to meet the growing complexities of life in a globalized world. AI algorithms and technologies have already made their way into everyday life, leading to improvement of human health, safety, and productivity. However, we need to balance these benefits with a careful deliberation of the unwanted side effects or even abuse of AI technologies. In compliance with ethical and moral principles, we need to ensure that AI systems benefit individuals and that AI's economic and social benefits are shared across society in a democratic and equal fashion.

If society approaches technological development primarily with fear and distrust, the technological progress will slow, and important steps toward ensuring the safety and reliability of AI technologies will be hindered. If society approaches technological innovation open-mindedly, technologies may have the potential to profoundly change society for the better. The basic building block to achieve this is trust. In order to foster and restore trust in technological advancement, we need to minimize risks, make systems verifiable, and build effective and accountable legislation, along with developing a new understanding of what trust in technologies can mean. The underlying psychological mechanisms of trust in human-technology relationships may extend the traditional trust concepts or reinvent the meaning of trust entirely. The process for this development has started already, but will require more practical experiences, experiments, and analyses in an open, discursive form with broad inclusion of societal stakeholders.

References

Ars Electronica (2019) Out of the Box – the Midlife Crisis of the Digital Revolution. *Ars Electronica Festival.* Linz, Austria, September 5th-9th 2019.

Bannister, F., & Connolly, R. (2011) Trust and transformational government: A proposed framework for research. *Government Information Quarterly*, 28, pp. 137-147. https://doi.org/10.1016/j.giq.2010.06.010

De Visser, E.J., Pak, R., & Shaw, T.H. (2018) From 'automation' to 'autonomy': the importance of trust repair in human–machine interaction. *Ergonomics*, 61(10), pp. 1409-1427. https://doi.org/10.1080/00140139.2018.1457725

Dietz, G., & Den Hartog, N.D. (2006) Measuring trust inside organisations. *Personnel Review,* 35 (5), pp. 557-588. https://doi.org/10.1108/00483480610682299

European Commission (2020) White Paper on Artificial Intelligence: a European approach to excellence and trust. Brussels, February 19th 2020. Available at: https://ec.europa.eu/info/sites/info/files/commission-white-paper-artificial-intelligence-feb2020_en.pdf   (accessed: March 3rd 2021)

Følstad A., Nordheim C.B., Bjørkli C.A. (2018) What Makes Users Trust a Chatbot for Customer Service? An Exploratory Interview Study. In: Bodrunova S. (eds) *Internet Science*. INSCI 2018. Lecture Notes in Computer Science, 11193, pp. 194-208. Springer, Cham. https://doi.org/10.1007/978-3-030-01437-7_16

Hancock, P.A., Kessler, T.T., Kaplan, A.D., Brill, J.C., & Szalma, J.L. (2020) Evolving Trust in Robots: Specification Through Sequential and Comparative Meta-Analyses. *Human Factors*, pp. 18720820922080-18720820922080. https://doi.org/10.1177/0018720820922080

Lankton, N.K., McKnight, D.H., & Tripp, J. (2015) Technology, Humanness, and Trust: Rethinking Trust in Technology. *Journal of the Association for Information Systems*, 16(10), pp. 880-918. https://doi.org/10.17705/1jais.00411

Mayer, R.C., Davis, J.H., & Schoorman, F.D. (1995) An Integrative Model Of Organizational Trust. *Academy of Management Review*, 20(3), pp. 709-734. https://doi.org/10.5465/amr.1995.9508080335

Ryan, M. (2020) In AI We Trust: Ethics, Artificial Intelligence, and Reliability. *Science and Engineering Ethics*, 26(5), pp. 2749-2767. https://doi.org/10.1007/s11948-020-00228-y

Schäfer, K.E., Chen, J.Y., Szalma, J.L., & Hancock, P.A. (2016) A Meta-Analysis of Factors Influencing the Development of Trust in Automation: Implications for Understanding Autonomy in Future Systems. *Human Factors*, 58(3), pp. 377-400. https://doi.org/10.1177/0018720816634228

Siau, K., & Wang, W. (2018) Building Trust in Artificial Intelligence, Machine Learning, and Robotics. *Cutter Business Technology Journal*, 31(2), pp. 47-53.

Toreini, E., Aitken, M., Coopamootoo, K., Elliott, K., Zelaya, C.G., & van Moorsel, A. (2020) The relationship between trust in AI and trustworthy machine learning technologies. In *Conference on Fairness, Accountability, and Transparency* (FAT* '20), January 27th –30th 2020, Barcelona, Spain. ACM, New York, NY, USA, pp. 272-283. https://doi.org/10.1145/3351095.3372834

Xu, F., Uszkoreit, H., Du, Y., Fan, W., Zhao, D., & Zhu, J. (2019) Explainable AI: A Brief Survey on History, Research Areas, Approaches and Challenges. In: Tang, J., Kan, M.Y., Zhao, D., Li, S., & Zan, H. (eds) Natural Language Processing and Chinese Computing. NLPCC 2019. *Lecture Notes in Computer Science*, (11839), pp. 563-574. Springer, Cham. https://doi.org/10.1007/978-3-030-32236-6_51

Weizenbaum, J. (1967). Contextual Understanding by Computers. *Communications of the ACM*, 10 (8), pp. 474-480. https://doi.org/10.1145/363534.363545

Part X
Learning from Crisis

Efficiency vs. Resilience: Lessons from COVID-19

Moshe Y. Vardi

Abstract Why was the world not ready for COVID-19, in spite of many warnings over the past 20 years of the high likelihood of a global pandemic? This chapter argues that the economic goal of efficiency, focused on short-term optimization, has distracted us from resilience, which is focused on long-term optimization. Computing also seems to have generally emphasized efficiency at the expense of resilience. But computing has discovered that resilience is enabled by redundancy and distributivity. These principles should be adopted by society in the "after-COVID" era.

By March 2020, COVID-19 (Coronavirus disease 2019) was spreading around the world. From a local epidemic that broke out in China in late 2019, the disease has turned into a raging pandemic, the likes of which the world has not seen since the 1918 Spanish flu pandemic. By then, thousands have already died, with the ultimate death toll growing into the millions. Attempting to mitigate the pandemic, individuals were curtailing travel, entertainment, and more, as well as exercising "social distancing," thus causing an economic slowdown. Businesses hoarded cash and cut spending in order to survive a slowdown of uncertain duration. These rational actions by individuals and businesses were pushing the global economy into a deep recession.

Observing the economic consequences of this unexpected crisis, William A. Galston asks in a March 2020 *Wall Street Journal* column[1] "What if the relentless pursuit of efficiency, which has dominated American business thinking for decades, has made the global economic system more vulnerable to shocks?" He went on to argue that there is a tradeoff between efficiency and resilience. "Efficiency comes through optimal adaptation to an existing environment," he argued, "while resilience requires the capacity to adapt to disruptive changes in the environment."

---

[1] https://www.wsj.com/articles/efficiency-isnt-the-only-economic-virtue-11583873155

M. Y. Vardi (✉)
Rice University, Houston, TX, USA
e-mail: vardi@cs.rice.edu

A similar point was made by Thomas Friedman in a May 2020 *New York Times* column:[2] "Over the past 20 years, we've been steadily removing man-made and natural buffers, redundancies, regulations and norms that provide resilience and protection when big systems—be they ecological, geopolitical or financial—get stressed. ... We've been recklessly removing these buffers out of an obsession with short-term efficiency and growth, or without thinking at all."

Both Galston and Friedman were pointing out that there is a tradeoff between short-term efficiency and long-term resilience. This tradeoff was also raised, in a different setting, by Adi Livnat and Christos Papadimitriou (2016). Computational experience has shown that simulated annealing, which is a local search—via a sequence of small mutations—for an optimal solution, is, in general, superior computationally to genetic algorithms, which mimic sexual reproduction and natural selection. Why then has nature chosen sexual reproduction as almost the exclusive reproduction mechanism in animals? Livnat and Papadimitriou's answer is that sex as an algorithm offers advantages other than good performance in terms of approximating the optimum solution. In particular, sexual reproduction favors genes that work well with a greater diversity of other genes, and this makes the species more adaptable to disruptive environmental changes, that is to say, more resilient.

The tradeoff between efficiency and resilience can thus be viewed as a tradeoff between short-term and long-term optimization. Nature seems to prefer long-term to short-term optimization, focusing on the survival of species. Indeed, Darwin supposedly said: "It's not the strongest of the species that survives, nor the most intelligent. It is the one that is most adaptable to change."

And yet, we have educated generations of computer scientists on the paradigm that analysis of algorithm only means analyzing their computational efficiency. As Wikipedia states:[3] "In computer science, the analysis of algorithms is the process of finding the computational complexity of algorithms—the amount of time, storage, or other resources needed to execute them." In other words, efficiency is the sole concern in the design of algorithms. (Of course, the algorithm has to meet its intended functionality.) *The Art of Computer Programming*,[4] a foundational text in computer science by Donald E. Knuth, is focused solely on efficiency. What about resilience? Quoting Galton again: "Creating resilient systems means thinking hard in advance about what could go wrong and incorporating effective countermeasures into designs." How can we make our algorithms more resilient?

Of course, fault tolerance has been part of the canon of computing-system building for decades. Jim Gray's 1998 Turing Award citation[5] refers to his invention of transactions as a mechanism to provide crash resilience to databases. Leslie Lamport's 2013 Turing Award citation[6] refers to his work on fault tolerance in

---

[2] https://www.nytimes.com/2020/05/30/opinion/sunday/coronavirus-globalization.html

[3] https://en.wikipedia.org/wiki/Analysis_of_algorithms

[4] https://en.wikipedia.org/wiki/The_Art_of_Computer_Programming

[5] https://amturing.acm.org/award_winners/gray_3649936.cfm

[6] https://amturing.acm.org/award_winners/lamport_1205376.cfm

distributed systems. Nevertheless, computer science has yet to fully internalize the idea that resilience, which to include reliability, robustness, and more, must be pushed down to the algorithmic level. Case in point is search-result ranking. Google's original ranking algorithm was PageRank,[7] which works by counting the number and quality of links to a page to determine how important the website is. But PageRank is not resilient to link manipulation, hence "search-engine optimization."

As pointed up by Friedman and Galston, the relentless pursuit of economic efficiency prevented us from investing in getting ready for a pandemic, in spite of many warnings over the past several years, and pushed us to develop a global supply chain that is quite far from being resilient. Does computer science have anything to say about the relentless pursuit of economic efficiency? Quite a lot, actually.

Economic efficiency means[8] that goods and factors of production are distributed or allocated to their most valuable uses and waste is eliminated or minimized. Free-market advocates argue[9] that through individual self-interest and freedom of production and consumption, economic efficiency is achieved and the best interests of society, as a whole, are fulfilled. But efficiency and optimality should not be conflated. The First Welfare Theorem,[10] a fundamental theorem in economics, states that under certain assumptions a market will tend toward a competitive, Pareto-optimal equilibrium; that is, economic efficiency is achieved. But how well does such an equilibrium serve the best interest of society?

In 1999, Elias Koutsoupias and Papadimitriou undertook (Koutsoupias and Papadimitriou 1999) to study the optimality of equilibria from a computational perspective. In the analysis of algorithms, we often compare the performance of two algorithms (e.g., optimal vs. approximate or offline vs. online) by studying the ratio of their outcomes. Koutsoupias and Papadimitriou applied this perspective to the study of equilibria. They studied systems in which non-cooperative agents share a common resource and proposed the ratio between the worst possible Nash equilibrium and the social optimum as a measure of the effectiveness of the system. This ratio has become known as the "Price of Anarchy,"[11] as it measures how far from optimal such non-cooperative systems can be. They showed that the price of anarchy can be arbitrarily high, depending on the complexity of the system. In other words, economic efficiency does not guarantee the best interests of society, as a whole, are fulfilled.

A few years later, Constantinos Daskalakis, Paul Goldberg, and Papadimitriou asked (Daskalakis et al. 2006) how long it takes until economic agents converge to an equilibrium. By studying the complexity of computing mixed Nash equilibria, they provide evidence that there are systems in which convergence to such equilibria can take an exceedingly long time. The implication of this result is that economic

---

[7] https://en.wikipedia.org/wiki/PageRank

[8] https://www.investopedia.com/terms/e/economic_efficiency.asp

[9] https://www.investopedia.com/terms/i/invisiblehand.asp

[10] https://en.wikipedia.org/wiki/Fundamental_theorems_of_welfare_economics

[11] https://en.wikipedia.org/wiki/Price_of_anarchy

systems are very unlikely ever to be in an equilibrium, because the underlying variables, such as prices, supply, and demand, are very likely to change while the systems are making their slow way toward convergence. In other words, economic equilibria, a central concept in economic theory, are mythical rather than real phenomena. This is not an argument against free markets, but it does oblige us to view them through a pragmatic, rather than ideological, lens.

Our digital infrastructure, which has become a key component of the economic system in developed countries, is one of the few components that did not buckle under the stress of COVID-19. Indeed, in March 2020, many sectors of our economy switched in haste to the WFH mode, "working from home." This work from home, teach from home, and learn from home were enabled (to an imperfect degree, in many cases) by the Internet. From its very roots of the ARPANET in the 1960s, resilience, enabled by seemingly inefficient distributivity and redundancy, was a prime design goal for the Internet (Yoo 2018). Resilience via distributivity and redundancy is one of the great principles of computer science that deserves more attention by the business community.

In summary, resilience is a fundamental, but under-appreciated, societal need. Both computing and economics need to increase their focus on resilience. It is important to note that markets and people tend to underprepare for low-probability or very long-term events. For example, car insurance is inefficient for insurance holder, though it offers resilience. Yet people are required to purchase car insurance, because many will not buy it otherwise. In other words, societal action is required to ensure resilience. It is important to remember this point, as many now argue that COVID-19 is just the "warm-up act" for the climate crisis.[12]

The big question is how the AC ("after-COVID") world will differ from the BC ("before-COVID") world. Fareed Zakaria wrote[13] in *The Washington Post* in October 2020: "The pandemic upended the present. But it's given us a chance to remake the future." Matt Simon wrote[14] in *Wired* in December 2020: "The COVID-19 pandemic has brought incalculable suffering and trauma. But it also offers ways for people—and even societies—to change for the better." I believe that resilience must be a key societal focus in the AC world.

---

[12] https://www.chronicle.com/article/covid-19-is-just-the-warm-up-act-for-climate-disaster

[13] https://www.washingtonpost.com/opinions/2020/10/06/fareed-zakaria-lessons-post-pandemic-world/

[14] https://www.wired.com/story/who-will-we-be-when-the-pandemic-is-over/

References

Daskalakis, C., Goldberg, P.W., and Papadimitriou, C.H. (2006) The complexity of computing a Nash equilibrium. STOC 2006: 71-78

Livnat, A. Papadimitriou, C.H. (2016) Sex as an algorithm: the theory of evolution under the lens of computation. Commun. ACM 59(11), pp. 84-93

Koutsoupias, E. and Papadimitriou, C.H. (1999) Worst-case Equilibria. Proc. 16th Annual Symposium on Theoretical Aspects of Computer Science, Lecture Notes in Computer Science 1563, Springer, pp. 404-413

Yoo, C.S. (2018) Paul Baran, Network Theory, and the Past, Present, and Future of the Internet. Colo. Tech. LJ 17, p.161.

Contact Tracing Apps: A Lesson in Societal Aspects of Technological Development

Walter Hötzendorfer

Abstract Overall, there might be more important aspects of the COVID-19 pandemic and the global fight against it than contact tracing apps. But the case of the contact tracing apps tells us an interesting story in the context of Digital Humanism. It shows us that the principle of privacy by design has reached practice and what it can mean and achieve. Unfortunately, it is also a lesson about the societal limitations of privacy by design in practice. It is a good thing that people are skeptical and ask questions about privacy and data protection. However, it is necessary to differentiate and try to make educated decisions or trust experts and civil society.

In the app stores of Google and Apple, we can find so many popular apps which track their users and trample on data protection. Most people do not question that at all and use these apps heedlessly. In the face of the COVID-19 pandemic, we have developed some of the most privacy-friendly and best-scrutinized apps, and people have questioned them widely—which is a good thing. In the resulting public discussion, it turned out to be difficult to explain a privacy-by-design solution to the public. Clearly, it is hard to understand how tracing of individual contacts and anonymity (or pseudonymity) can be possible at the same time.

One particularly perfidious characteristic of SARS-CoV-2 is that an infected person can already infect others while she still feels perfectly healthy. Therefore, if Alice has been infected by Bob, Alice must be warned immediately and stay at home as soon as Bob learns that he has been infected. Then, those who Alice would have otherwise met later are saved. Contact tracing is a measure to achieve such warnings of contact persons.

Almost immediately after we realized that the virus had reached Europe in early 2020, various projects started in different countries to develop an app which could supplement the usual contact tracing carried out by health authorities. Out of these developments, soon a broad and qualified international discussion emerged about

W. Hötzendorfer (✉)
Research Institute AG & Co KG, Vienna, Austria
e-mail: walter.hoetzendorfer@researchinstitute.at

different concepts on how to implement app-based contact tracing. Surprisingly, it turned out that these concepts, when optimized for effective warning, or when optimized for privacy, or when optimized for ease of implementation,[1] converge to the same result.[2] In other words, contact tracing apps are a rare case where there is an optimum solution when considered by these three criteria.[3] This approach is the Bluetooth-based creation of an anonymous contact diary kept locally on the smartphone, later called the "decentralized approach," which was particularly well elaborated and documented under the designation Decentralized Privacy-Preserving Proximity Tracing (DP-3T).[4] After intensive public discussion[5]—which unfortunately is usually anything but granted—this factually best solution prevailed in practice in most countries. This is opposed by a "centralized approach," where pseudonymous records of contacts, at least of the infected users, are transmitted to a server. In this architecture, it cannot be ruled out that a malicious server operator or attacker can infer a social graph or even an interaction graph of participating persons (Troncoso et al. 2020, pp. 43 f). The example shows how important fundamental architectural decisions are for the privacy of the users (Hötzendorfer 2020). There are many more details of applied privacy by design in DP-3T and similar concepts. See Troncoso et al. (2020) for the details.

Soon after the first apps were developed, it became clear that without intervention on the operating system level, Bluetooth could not be used in the required way, especially when the app should run in the background. Then something historic happened: Google and Apple teamed up to implement a privacy-by-design concept developed by European researchers (among others). They implemented a concept for Bluetooth-based contact tracing very similar to DP-3T into Android and iOS, respectively.[6] Of course, it is a problem in itself that these companies are so

---

[1] It must be noted that ease of implementation not only shortens time to market but is also an important security feature. The more complex a system is, the more difficult it is to secure, and this relation is definitely not linear.

[2] For example, a Bluetooth-based solution is both temporally and geographically more precise and more privacy-friendly than a GPS-based solution. It would go beyond the scope to explain all the details here; see Troncoso et al. (2020).

[3] Admittedly, there may be additional parameters, but it is apparent that these three parameters are pivotal for bringing an effective, lawful, and trusted app into the field as quickly as possible.

[4] See https://github.com/DP-3T.

[5] See, for example, Contact Tracing Joint Statement (available at https://www.esat.kuleuven.be/cosic/sites/contact-tracing-joint-statement/); CCC, "10 Prüfsteine für die Beurteilung von 'Contact Tracing'-Apps" (available at https://www.ccc.de/de/updates/2020/contact-tracing-requirements); European Data Protection Board, "Guidelines 04/2020 on the use of location data and contact tracing tools in the context of the COVID-19 outbreak" (available at https://edpb.europa.eu/our-work-tools/our-documents/guidelines/guidelines-042020-use-location-data-and-contact-tracing_en); Bayer et al., "Technical and Legal Review of the Stopp Corona App by the Austrian Red Cross" (available at https://epicenter.works/document/2497); and Neumann, "'Corona-Apps': Sinn und Unsinn von Tracking" (available at https://linus-neumann.de/2020/03/corona-apps-sinn-und-unsinn-von-tracking).

[6] https://covid19.apple.com/contacttracing

dominant and powerful that the world is practically unable to implement such a system without their good will (Veale 2020).

However, Google and Apple may not have chosen the decentralized approach out of noble privacy considerations. In light of the potential information power that comes with the centralized approach, one can raise the hypothesis that they did not want to decide which governments are trustworthy enough to give them these powers and which ones are not. In any case, they implemented the decentralized approach, and so the most privacy-friendly solution prevailed in practice.

As many people may not know, the GDPR, which ushered in a new era of data protection when it came into force in 2018, has not significantly changed the substantive data protection law in Europe. Rather, its fundamental impact results from the penalties it imposes for conduct that, for the most part, was already unlawful before and from the momentum and focused public discussion it created all over Europe and beyond. However, the GDPR did introduce a new principle: data protection by design. This fundamental principle wants us to build privacy into the design of systems from the start. Only slowly an understanding is being developed of what this requirement means in practice and how it can be systematically fulfilled. DP-3T and related concepts and their implementation by Apple and Google can be seen as one of the first widespread real privacy-by-design solutions in the sense that it demonstrated that with a privacy-first attitude, the key functional requirements of a software can be fulfilled without compromise.

However, it made us realize that we are not there yet. The step we took here, from having a sound body of data protection law which should theoretically protect users installing an app to having the app implemented as a data protection by design solution and making that transparent in every detail, was not enough to gain the trust of the users.

This is particularly noteworthy since the quality and depth of the public technological discussion was remarkable. For example, in Austria, not only the Data Protection Authority but also the broader data protection community was involved in the development of the app very early, and the nationally and internationally recognized NGOs epicenter.works and NOYB and the information security research center SBA Research carried out a technical and legal review and published a report containing a list of recommendations which were immediately implemented.[7] The important realization, that we must actively participate in shaping technology if we are to exercise political control over it, seemed to have suddenly taken effect in civil society and the scientific community. The European Data Protection Supervisor stated: "The public discussion about specific privacy features for a new application, which was only in the early phases of development, was a completely new phenomenon."[8]

---

[7] https://epicenter.works/document/2497

[8] https://edps.europa.eu/press-publications/press-news/blog/what-does-covid-19-reveal-about-our-privacy-engineering_en

At least from an Austrian and German perspective, it seems that although independent, renowned privacy experts and activists gave a good verdict on the app and published their reasons in detail, they were not able to change the public opinion significantly. Of course, it is almost impossible to check whether the superior concept was implemented correctly in every detail, and bugs can never be ruled out. Also, there are still some factual arguments against contact tracing apps, such as the doubts about the suitability of Bluetooth.[9] However, many people seem to continue to refuse the app primarily out of a privacy-related gut feeling, even though the privacy of the app had already been thoroughly tested by independent experts and the results were openly available in detail. Therefore, the question arises, how can we as technological experts replace gut-based opinions in the population with fact-based science-driven opinions in the face of the following insight: Many mechanisms and measures that make a technology more secure or less privacy-intrusive are complicated. Take encryption as the most obvious example. In many cases, the exact same mechanism that makes us recommend a software makes it difficult to explain to the general public why we can do so.

Unfortunately, this effect cannot be simply explained by differences in the level of education and expertise in society or even by the increasing hostility toward science. Doubts concerning the app based on gut feelings without substantive arguments were also spread publicly by scholars from other disciplines as well as privacy advocates who had not carried out a thorough analysis of the app themselves. This is not to argue, in a blunt and dangerous way, anyone who has doubts about the app has simply no clue about it, and of course, skepticism about technological developments and specific systems in particular is always appropriate. But if a lot of detailed privacy analysis based on scientific methods is publicly available, this cannot be ignored in the public discussion. If intellectuals and experts in non-technical disciplines write or speak publicly about privacy aspects of a technological system such as a contact tracing app, they should be aware of the existence of such widespread scientific evidence and base their discussion on it, the same way in which they base their discussion on the broad consensus among epidemiologists.

This is not to say that the lack of widespread use of contact tracing apps can only be attributed to factually unfounded privacy concerns. Another problem of contact tracing apps is obviously that their use is a rather passive experience. The app not doing anything recognizable during normal use could lead the user to believe that it is not working.[10] But in any case, experts were not able to convince a broad public

---

[9] See, e.g., Schneier, "Me on COVID-19 Contact Tracing Apps" (available at https://www.schneier.com/blog/archives/2020/05/me_on_covad-19_.html).

[10] I am convinced that the number of active users of contact tracing apps could have easily been boosted. One option would be to integrate other functionalities into the app, such as the extension of the contact tracing functionality by an (e.g., QR code based) anonymous check-in option for meetings, meeting rooms, restaurants, and other kinds of locations. There are developments in this direction in some countries. Another option would be a lottery, i.e., to announce that one random user per week will be informed by a push notification that she won 1000 or be it 10,000 euros—a rather cheap measure compared to the cost of the pandemic (cf. https://logbuch-

that the decentralized approach and the apps based on it are in fact harmless in terms of privacy and data protection. The European Data Protection Supervisor concludes that: "From all reactions, it appears that the biggest inhibitor to wide uptake and use of tracing apps is the lack of trust in their confidentiality."[11]

Has the world become so complicated that a broad majority cannot take qualified (democratic) decisions concerning a growing number of domains? One way forward is to strengthen trust in experts and science. But to be honest, we have to realize that this cannot fully succeed in explaining privacy-by-design solutions to a broader public. This might be related to the fact that privacy by design is in a way an attempt to control technology with more technology. And this at least makes things more complicated and hence more complicated to explain.

Clearly, there are domains where digital solutions are conceptually completely inappropriate, and the paper-based solution fulfils the essential requirements appropriately, e.g., a secret ballot.[12] But the domain of contact tracing apps is a good example where only technology enables a suitable solution, i.e., tracing and anonymity at the same time, which is conceptually impossible to achieve with any paper-based approach.

In many other domains, we might not be able to find such elegant privacy-by-design solutions that fulfil all functional requirements as DP-3T does here. As I am writing these lines exactly 1 year after the Austrian Stopp Corona App was released, another app-based "solution" in the context of the pandemic is around the corner: the "green" app-based pass for demonstrating the fact of being tested, vaccinated, or immune due to a past infection. Unfortunately, here the perfect privacy-by-design concept for implementing such a system does not (yet) suggest itself. At the same time, this is a much more crucial domain than contact tracing because people will be under much more factual pressure to use such a system if they want to participate in public life again. However, it seems that wherever the "green pass" is discussed, it is much clearer than it was a year ago that such a solution must meet the highest standards in privacy and data protection.

To conclude, I think this is the positive legacy of the contact tracing apps in the context of Digital Humanism: We can expect that applied privacy by design will become more common. Also, the public debate about privacy and data protection was elevated onto a new level. Mankind needs to find ways to actively shape technological progress for the greater good, and therefore civil society and the scientific community must involve themselves as it happened here. However, we

---

netzpolitik.de/lnp385-fuempf-blockchains). Experience tells me that fatally, for many people, this would very quickly wipe away their vague privacy concerns.

[11] https://edps.europa.eu/press-publications/press-news/blog/what-does-covid-19-reveal-about-our-privacy-engineering_en

[12] Among several reasons, this is mainly due to the lack of comprehensibility of a digital election process that meets the requirements of a secret and secure election. Therefore, electronic voting would suffer enormously under the same problem which is identified as a key issue in this contribution, the lack of trust in technological solutions and in the experts who are able to understand them, which is why electronic voting is particularly worth mentioning in this context.

also learned that this is not enough: As technological development is making the world more difficult to understand every day, we need to find ways to explain "good" technology to the people, including intellectuals and experts in other fields, while maintaining a sound and productive skepticism toward technological developments that influence our lives.

References

Hötzendorfer, W. (2020) 'Zum Verhältnis von Recht und Technik: Rechtsdurchsetzung durch Technikgestaltung' in Hötzendorfer, W., Tschohl, C., Kummer, F. (eds.) *International Trends in Legal Informatics, Festschrift for Erich Schweighofer.* Bern: Editions Weblaw, 419–437.
Troncoso C. et al. (2020) 'Decentralized Privacy-Preserving Proximity Tracing', Whitepaper, DP-3T Consortium, 25 May. Available at: arXiv:2005.12273
Veale, M. (2020) 'Privacy is not the problem with the Apple-Google contact-tracing toolkit', The Guardian, 1 July. Available at: https://www.theguardian.com/commentisfree/2020/jul/01/apple-google-contact-tracing-app-tech-giant-digital-rights.

Data, Models, and Decisions: How We Can Shape Our World by Not Predicting the Future

Niki Popper

Abstract Modelling and simulation can be used for different goals and purposes. Prediction is only one of them, and, as this chapter highlights, it might not be the main goal—even if it was in the spotlight during the COVID-19 crisis. Predicting the future is a vanity. Instead, we aim to prevent certain events in the future by describing scenarios, or, even better, we try to actively shape the future according to our social, technological, or economic goals. Thus, modellers can contribute to debate and social discourse; this is one of the aims of Digital Humanism.

"I don't try to describe the future. I try to prevent it." This Ray Bradbury quote from 1977 was cited by Theodore Sturgeon in Emtsev and Parnov (1977, p. viii): "In a discussion of (Orwell's) book 1984, Bradbury pointed out that the world George Orwell described has little likelihood of coming about—largely because Orwell described it. 'The function of science fiction is not *(only)* to predict the future, but to prevent it.' Bradbury said."[1]

The use of modern simulation methods also often falls prey to the misunderstanding that prediction is its main goal. In my opinion, it is not our purpose to predict the future. Instead, we aim to prevent certain events in the future by describing scenarios or—even better—try to actively shape the future according to our social, technological, or economic goals. We can thereby contribute to discussions and social discourse; this is one of the intentions of Digital Humanism. One of the most important scientific contributions to achieving this aim has been the development of innumerable types of models that are fed with all kinds of data. This wealth rather complicates things...

---

[1] Theodore Sturgeon wrote this in the preface to a Russian science fiction book. One can, of course, concede that the idea of preventing disaster was probably more prevalent during the Cold War era than that of bringing about positive change, as compared to recent decades.

N. Popper (✉)
Vienna University of Technology, Vienna, Austria
e-mail: nikolas.popper@tuwien.ac.at

H. Werthner et al. (eds.), *Perspectives on Digital Humanism*,
https://doi.org/10.1007/978-3-030-86144-5_40

At first glance, the notions of shaping the future and of preventing undesired events from happening do not differ from each other. They both revolve around the fact that we usually want to not only predict the future but to actually change it (for the better). Some kinds of models, like the weather forecast, focus on prediction. We mostly just want to know how likely it is that we will need an umbrella and not why. Thus, the design of those kinds of models differs greatly from the ones that will be described in this chapter, i.e., those being used by my group in the ongoing COVID-19 crisis. These are models that can show different possible outcomes for different decisions and are used to support the discussion of the available variety of strategies.

In January 2020, my group at TU Wien began applying our model of the virtual Austrian population, its interactions and connections to infrastructure, measures, and policies in order to model the COVID-19 crisis (Bicher et al. 2021a). We are now able to map different aspects of the COVID-19 crisis that interact with each other on an individual level. These are, among others, strategic aspects such as (A) setting and cancelling measures; (B) testing, screening, and isolation strategies; (C) vaccination; and (D) the development of new therapeutic concepts. From a systemic perspective, it is possible to implement (E) changes in viral properties (such as mutations), (F) changes in the population (e.g., through natural immunization or vaccination), or (G) changes in environmental influences.

We have contributed our work to the Austrian government since March 2020. Initially, this support took the form of short-term predictions together with other research groups as joint action in the Austrian Prognosis Consortium (Bicher et al. 2021b). Subsequently, in addition, we started to communicate the relationships between measures, dynamics, decisions, and social and epidemiological outcomes. The main approaches have been vaccination programs (Jahn et al. 2021) and screening strategies and how they can shape society for better or worse.

During the COVID-19 crisis, we have learned that there are no technological solutions without integrating peoples' needs, weaknesses, hopes, and ambitions. Simulation, models, and decision tools have to be integrated into processes based on the foundations of Digital Humanism. We also need solutions in order to cope with the lack of transparent and reliable data that can be used for our models in accordance with the European Data Protection Regulation (GDPR).

Bradbury's statement refers to the fact that prediction can prevent a thing because we have become aware of it and take countermeasures. That is something where modelling can contribute. Modelling and simulation have to do more than sketch an "outcome," as science fiction does so well. Ideally, we also want to describe feasible ways to improve something in the future. To be able to do so, we need to understand interrelationships and describe causalities without ever leaving the safe ground of steady data. This is the fundamental concept of "decision support" as sought by politicians, managers, and others with the decision-taking powers.

Do we need accurate predictions in order to generate reliable decision support? Not necessarily. In fact, prediction can even be a hindrance in the process of change because it reinforces the impression that the future is already decided. Might

Laplace's demon[2] (Laplace 1825, p. 4) have long been refuted already? While it is necessary to think and work in scenarios, we then also need to integrate these thoughts into the change process. Models can only be one piece of the puzzle of "decision support." They need to be embedded into the bigger picture, i.e., other processes.

In my experience, it is unavoidable to make use of a range of different established methodological approaches and, moreover, combine these to create tailor-made processes which make it possible to link the respective advantages and disadvantages of each method, laying the groundwork for "better decisions." I like to think of it as a gradual process with feedback loops:

(1) **Get your data straight.** In their first step, modellers collect and analyze data. Hypotheses are generated on the basis of data with the help of different methods, such as statistics or artificial intelligence. This approach makes it possible to make forecasts. During the COVID-19 crisis, early models allowed us to make basic statements about the current situation or to compare international developments. However, these concepts often mislead us into believing that we can continue to extrapolate developments. Moreover, we are still lacking valid and quality-assured data.

(2) **Establish correlations and causalities and describe relationships.** In this step, macroscopic models are used to couple the formalized data with causal or relational hypotheses. Examples of this approach have existed for many decades. Earliest works came from Norbert Wiener (Wiener 1961); Jay W. Forrester was a pioneer (Forrester 1973). System dynamics is one representative of the linkage between model representation through differential equations and the modelling process with non-mathematicians. *The Limits to Growth. A Report for the Club of Rome's Project on the Predicament of Mankind* by Meadows (Meadows et al. 1972) was an early example of the impact such approaches can have. It has enabled us to use system dynamics in order to develop our understanding of feedback loops and regulated systems in economics and ecology. With approaches such as system dynamics and differential equations, we can describe relationships and explain, for example, exponential behavior and logistic growth. While these aspects have particularly come to the fore in the COVID-19 crisis, they have in fact been helping us to better understand the mechanisms of actions in therapy analysis for many years. Causal analysis has further facilitated this understanding since 2000 (Pearl 2000) by addressing issues

---

[2]Laplace wrote: "We ought then to regard the present state of the universe as the effect of its anterior state and as the cause of the one which is to follow. Given for one instant an intelligence which could comprehend all the forces by which nature is animated and the respective situation of the beings who compose it—an intelligence sufficiently vast to submit these data to analysis—it would be embrace in the same formula the movements of the greatest bodies of the universe and those of the lightest atom; for it, nothing would be uncertain and the future, as the past, would be present to its eyes."

where traditional regression analyses fail in combination with time-varying confounding.[3]

(3) **Describe emergent behavior.** Finally, modelling with agents and networks makes it possible to describe dynamic socio-technical systems, such as when health system interventions are implemented in a heterogeneous population. Models try to use hypotheses and data in order to represent a given process both at the individual level and at the level of the real infrastructure. One example of this is the COVID-19 model developed by our group (Bicher et al. 2021a). These methods allow us to describe emergent behavior as well as frequently analyze and pre-estimate counter-intuitive behavior.

By being both complementary to and based upon each other, these three approaches to predicting the future contribute to an advanced understanding of the world as it is. They have enabled us to model, among other things, concrete decision support for vaccination prioritization (Jahn et al. 2021), screening strategies, as well as an assessment of ongoing immunization and its impact on spread dynamics (Rippinger et al. 2021).

Criticism of all three approaches is necessary and justified, but cannot be discussed in detail here. Very generally, data-driven models very easily trigger the hope for predictions. Causal models are often criticized for the lack of available means of validation. In the past, it has been almost impossible to link these models with data. COVID-19 was a global situation that revealed possibilities as well as weaknesses.

Sensible decision support for humans will in the future hinge on a combination of all these approaches. This is a key realization. Agent-based models are particularly well suited for integrating new data structures on an ongoing basis. During the COVID-19 crisis, this included epidemiological data, vaccination data, mobility data, information on home office use, school closures, weather data, and many more. The effort it takes to verify, validate, calibrate, and consider their reproducibility is enormous, but it is worth it. In order to really achieve the desired results, these agent-based models need to be combined with data-driven and macroscopic approaches. In Austria, the COVID-19 crisis inspired the formation of a consortium early on, so that all approaches could be compared (Bicher et al. 2021b). Together, we cover a range of methods to be able to answer different research questions.

One of the tasks and aspects of modelling is, of course, forecasting. The models that are good at forecasting are, however, different from those that are used for strategy development. Each question needs an appropriate model and every model needs the right questions. We have now come full circle: we can output the corresponding data from agent-based models as synthetic data sets (Popper et al. 2021) in order to examine the very models we have used with statistical methods.

---

[3]This means that there are variables that are simultaneously confounders (common causes of treatment and outcome) and intermediate steps, i.e., on the causal chain treatment that leads to an outcome. In other words: Confounders are also affected by treatment. It is difficult to determine when one has identified "sufficient" causal chains.

The gathering and handling of raw data, including a true understanding of the patterns and relationships therein, are undoubtedly vital to establish a solid basis for models. However, when a model aims to provide decision support, we must also be able to identify, represent, and reproduce the dynamics of a system—i.e., the behavior of the population. That is what makes it possible to not only predict future events but actually understand their underlying reasons.

Like other philosophical problems, Laplace's demon will continue to occupy our thoughts. We might even want this particular demon to act as our sparring partner. We will continue to ponder what purpose models can serve exactly and how stochasticity plays out in different models and "if-then" predictions. Also, we might want to have a closer look at those who gather the data, finding correlations and describing emergent behaviors.

In the end, we have to be aware of our limits and always keep in mind what our form of decision-taking support can actually contribute:

A few years ago, I presented an early version of the agent-based network model that we are currently using in the context of COVID-19 at a major meeting on the issue of influenza vaccination strategy. A medical professional approached me after the presentation and asked triumphantly: "Can you tell me now how many patients with the flu we can expect to present next year on the 17th of March?". My answer was: "No, I can't—you've got me there! But that knowledge would be utterly pointless, anyway. What we can do with our model, however, is to tell you which strategy you can use in order to minimise the number of infected people, or, indeed, to produce maximum damage."

References

Bicher M., Rippinger C., Urach C., Brunmeir D., Siebert U., Popper N. (2021a) Evaluation of Contact-Tracing Policies Against the Spread of SARS-CoV-2 in Austria – An Agent-Based Simulation accepted in Medical Decision Making, https://doi.org/10.1101/2020.05.12. 20098970

Bicher M., Zuba M., Rainer L., Bachner F., Rippinger C., Ostermann H., Popper N., Thurner N., Klimek P. (2021b) Supporting Austria through the COVID-19 Epidemics with a Forecast-Based Early Warning System, medRxiv 2020.10.18.20214767; https://doi.org/10.1101/2020.10.18. 20214767

Emtsev M. and Parnov E. (1977) World Soul, Introduction by Theodore Sturgeon, page viii, translated from Russian by Antonina W. Bouis, Macmillan Publishing Co., New York, *researched at* https://quoteinvestigator.com/2010/10/19/prevent-the-future/

Forrester J.W., (1973) World Dynamics. Cambridge, Mass.: Wright-Allen Press.

Jahn B., Sroczynski G., Bicher M., Rippinger C., Mühlberger N., Santamaria J., Urach C., Schomaker M., Stojkov I., Schmid D., Weiss G., Wiedermann U., Redlberger-Fritz M., Druml C., Kretzschmar M., Paulke-Korinek M., Ostermann H., Czasch C., Endel G., Bock W., Popper N. and Siebert U. (2021) "Targeted COVID-19 Vaccination (TAV-COVID) Considering Limited Vaccination Capacities—An Agent-Based Modeling Evaluation," Vaccines, vol. 9, no. 5, p. 434, Apr. 2021, https://doi.org/10.3390/vaccines9050434

Laplace P.S. (1825) Essai philosophique sur les probabilités , A Philosophical Essay on Probabilities translated from French by Frederick Wilson Truscott and Frederick Lincoln Emory, John

Wiley & Son, Chapman & Hall, 1902, https://bayes.wustl.edu/Manual/laplace_A_philosophi cal_essay_on_probabilities.pdf

Meadows, Donella H; Meadows, Dennis L; Randers, Jørgen; Behrens III, William W (1972) The Limits to Growth; A Report for the Club of Rome's Project on the Predicament of Mankind. New York: Universe Books. ISBN 0876631650. Retrieved 26 November 2017.

Pearl J. (2000) Causality. Cambridge university press.

Popper N., Zechmeister M., Brunmeir D., Rippinger C., Weibrecht N., Urach C., Bicher M., Schneckenreither G., Rauber A. (2021) Synthetic Reproduction and Augmentation of COVID-19 Case Reporting Data by Agent-Based Simulation, Data Science Journal 20(1), https://doi.org/10.5334/dsj-2021-016

Rippinger, C., Bicher, M., Urach, C. et al. (2021) Evaluation of undetected cases during the COVID-19 epidemic in Austria. BMC Infect Dis 21, 70, https://doi.org/10.1186/s12879-020-05737-6

Wiener N. (1961) Cybernetics: Or Control and Communication in the Animal and the Machine. Paris, (Hermann & Cie) & Camb. Mass. (MIT Press) ISBN 978-0-262-73009-9; 1948, 2nd revised ed. 1961.

Lessons Learned from the COVID-19 Pandemic

Alfonso Fuggetta

Abstract The COVID-19 pandemic is having a tragic and profound impact on our planet. Thousands of lives have been lost, millions of jobs have been destroyed, and the life of billions of people has been changed and disrupted. In this dramatic turmoil, digital technologies have been playing an essential role. The Internet and all its services have enabled our societies to keep working and operating; social networks have provided valuable channels to disseminate information and kept people connected despite lockdowns and the block of most travels; most importantly, digital technologies are key to support researchers, epidemiologists, and public officers in studying, monitoring, controlling, and managing this unprecedented emergency.

After more than a year, it is possible and worthwhile to propose some reflections on the strengths and weaknesses we have experienced and, most importantly, on the lessons learned that must drive our future policies and roadmaps. This is unavoidable not just to improve our ability to react to these dramatic situation, but, most importantly, to proactively design and develop a better future for our society.

1 Lesson 1: Increase Digitalization

At the beginning of March 2020, as many other countries, all of Italy was put on lockdown: stores and schools were closed, most workers started working remotely, industries were forced to operate with limited staff, and traveling was basically cancelled. The key infrastructure that kept the country alive and operating was the Internet. Nevertheless, there were (and still are) problems that could not be solved instantaneously: significant portions of the territory do not have adequate network access and services; most public administrations do not manage all the information digitally and, most important, operate as separate silos that do not exchange and integrate their data and processes; too many companies were unprepared to work as

A. Fuggetta (✉)
Cefriel – Politecnico di Milano, Milan, Italy
e-mail: alfonso.fuggetta@cefriel.com

H. Werthner et al. (eds.), *Perspectives on Digital Humanism*,
https://doi.org/10.1007/978-3-030-86144-5_41

an agile and decentralized organization and to implement smart-working policies and procedures; and the health system was overwhelmed and had to deal with a sort of war-time emergency. The key problem was the *lack of preparation*: we didn't plan, and we didn't organize our society to deal with these kinds of events. We just reacted, and reaction cannot instantaneously fix problems and issues of such magnitude.

This was not just Italy's problem. All countries have been struck and caught by surprise. The first lesson we should learn from this emergency is that optical fibers cannot be laid down "on demand," processes cannot be set up "when needed," the society cannot adapt "by decree," and people and companies cannot just "react": we need to *anticipate and shape the future rather than just dealing with it when emergencies occur*.

2 Lesson 2: Speed Is More Important Than Money

When the crisis erupted, Italy was unable to procure all the masks needed to deal with the emergency. It was not a matter of a lack of resources: there were no production and sourcing capabilities, and it took weeks and months to create them. A country with a very high GDP, a member of the G20, was unable to procure basic devices such as masks. Similarly, when Italy decided to launch a contact tracing app, the real issue was not the cost of developing the app or integrating it with the national health information system: the real issue was the time needed to put it into operation. As an additional example, the same problem occurred to define and deploy an interoperability strategy for the different COVID-19 apps developed in Europe.

In an emergency, *the critical resource is time, not money*. Consequently, the speed and efficiency of processes are of paramount importance. It depends on the level of preparedness (as discussed in the previous point) and on the efficiency and clarity of the command chain within the different branches of the government and of the society in general.

Time cannot be procured.

3 Lesson 3: We Need to Find a New Balance Between Privacy and Public Good

The need to trace and control the pandemic has generated a heated and turbulent discussion about a crucial and sensitive issue: privacy preservation. Indeed, the characteristics of the virus do require the availability of fast and pervasive mechanisms to identify infected people that must be isolated so that the infection is blocked as soon as possible. This process can be enabled and accelerated by digital technologies and processes that trace and exploit a number of personal data and information.

In this respect, we have seen different approaches that have explored the entire spectrum of possibilities, from light access to a minimum of personal data to a pervasive and intrusive penetration into the private lives of citizens. Too often, the debate on this topic has been unable to strike a reasonable balance between these different tensions nor to instill enough trust in the public opinion.

We need new rules, policies, and mechanisms that are able to find the *appropriate balance and trade-off between promotion of societal interests and protection of freedom and civil rights*. This has to be achieved by exploiting two key directions. First, as indicated by the EU Digital Services Act and related legislative initiatives,[1] it is vital to provide transparent mechanisms to inform citizens about the policies and rules used to collect and use their personal data and to protect their rights and privacy. Second, it is crucial and urgent to increase our investments in education to ensure that every European citizen uses digital technologies in an informed and knowledgeable way.

4 Lesson 4: Interoperability Is Vital

In Italy as well as across European countries, one of the most critical problems has been the integration of different information and data sources that were key to monitor and control the pandemic. Two are the main problems:

1. Technical interoperability standards and enabling infrastructures
2. Common data schema and semantic models to interpret and exploit data coherently and effectively

Addressing these issues requires an incredible effort, similar in nature—but more challenging as far as its scope and complexity are concerned—to the creation of the GSM standard and related infrastructure. In Europe, there are two important initiatives that are trying to tackle these challenges: the Gaia-X project[2] and the International Data Spaces Association (IDSA).[3] The former aims at defining a European industrial and market strategy to promote and exploit cloud computing and related services; the latter aims at creating exchange standards for critical data assets.

As for the development of GSM, the Internet, and other digital technologies, standardization is an essential and crucial foundation to promote innovation, growth, and societal impact. The key concept here is *coopetition*: it is vital to *cooperate* in the creation of standards and common enabling infrastructures that define a level-playing field while *competing* to offer the best services to citizens and companies based on the availability of these enabling assets.

---

[1] Directorate-General CONNECT of the European Commission. "The Digital Services Act." https://digital-strategy.ec.europa.eu/en/policies/digital-services-act-package

[2] https://www.data-infrastructure.eu/

[3] https://internationaldataspaces.org/

5 Lesson 5: Half the Number of Words, Double the Quality of Communication

One of the major failures that we have experienced in this pandemic is the inability to provide clear and convincing messages to citizens. Such an incredible crisis would have demanded for an evidence-based communication strategy that too often has been completely absent or insufficient. Even worse, we have been overwhelmed by an unmanageable amount of incoherent, confusing, and often contradicting messages.

Government, companies (e.g., big pharma), and scientific institutions need to *raise the bar in their communication strategies and practices*, to provide the public with appropriate and trustworthy information on the evolution and management of the emergency. This has to be achieved by combining clear and coordinated communication strategies and procedures (especially as far as public bodies are concerned) with a streamlined, timely, and coherent exploitation of digital media and social network.

6 Lesson 6: Competences Are the Priority

Managing complexity demands for a substantial improvement of our competences as public officers, researchers, educators, and citizens. We cannot manage such a challenging crisis if we do not have the cultural and scientific means that can help us understand and deal with it. Of course, different roles would require different levels of knowledge and expertise. However, as a society, we need to elevate the average level of education and competence of each citizen. This requires a coordinated set of actions:

1. Provide digital awareness courses at all levels of the education system and with a lifelong learning vision.
2. Rethink and reimagine all curricula and disciplines to keep into account how digital technologies have dramatically changed them (e.g., geography and math).
3. Provide support and guidance to our students in the selection of their studies and careers so that we can increase the number of professionals in STEM.
4. Establish or strengthen curricula that bring together technology and humanities.
5. Fight school drop-out and ensure good levels of education for every citizen.

7   Lesson 7: Digital Technologies and Infrastructures Are Key European Security Affairs

Apple and Google played a key role in the development of contact tracing apps. Even if—unfortunately—they didn't have a significant and wide impact on the containment of the epidemic, as Europeans, we have been basically dependent on the decisions and strategies of foreign industries. Similarly, our working and social activities heavily rely on technologies developed by American and Asian companies. It would be silly to simple promote a simplistic and unfeasible technological protectionism. At the same time, as the handling of the vaccine production and procurement processes has demonstrated, the European Union needs to define a strategy that considers science and high-tech policy a strategic and security affair, and not just a cultural or economic issue.

The Need for Respectful Technologies: Going Beyond Privacy

Elissa M. Redmiles

Abstract Digital technologies, the data they collect, and the ways in which that data is used increasingly effect our psychological, social, economic, medical, and safety-related well-being. While technology can be used to improve our well-being on all of these axes, it can also perpetrate harm. Prior research has focused near exclusively on privacy as a primary harm. Yet, privacy is only one of the many considerations that users have when adopting a technology. In this chapter, I use the case study of COVID-19 apps to argue that this reductionist view on technology harm has prevented effective adoption of beneficial technology. Further, a privacy-only focus risks perpetuating and magnifying existing technology-related inequities. To realize the potential of well-being technology, we need to create technologies that are respectful not only of user privacy but of users' expectations for their technology use and the context in which that use takes place.

Digital technologies are increasingly intertwined with lived experiences of well-being. The ways in which we use technologies, and the ways in which technologies use our data, affect our psychological, social, economic, medical, and safety-related well-being. For example, being able to check in on the well-being of others during natural disasters can bolster the strength of our physical-world communities and enhance our personal feelings of safety (Redmiles et al. 2019). In the health space, there is growing excitement and promising evidence for prescribing technologies to aid in the management of chronic illness (Byambasuren et al. 2018).

Despite their potential to improve our well-being, these same technologies can also perpetrate harm. Much of the dialogue regarding the technological harms of well-being technologies focuses specifically on data privacy risks: how the misuse of user data can create psychological, social, economic, or safety-related harms (Vitak et al. 2018; Redmiles et al. 2019).

E. M. Redmiles (✉)
Max Planck Institute for Software Systems, Saarbrücken, Germany
e-mail: eredmiles@mpi-sws.org

© The Author(s) 2022
H. Werthner et al. (eds.), *Perspectives on Digital Humanism*,
https://doi.org/10.1007/978-3-030-86144-5_42

Privacy has been shown to be a key, and growing, concern for users when considering whether to adopt new technologies, including well-being-related technologies. However, privacy is far from the only consideration that effects whether a user will adopt a new technology. Here, I argue that we have developed a reductionist focus on privacy in considering whether people will adopt a new technology. This focus has prevented us from effectively achieving adoption of beneficial technologies and risks perpetuating and magnifying inequities in technology access, use, and harms.

By focusing exclusively on data privacy, we fail to fully capture user's desire for *respectful technologies*: systems that respect a user's expectations for how their data will be used and a user's expectations for how the system will influence their life and the contexts surrounding them. I argue that user's decisions to adopt a new technology are driven by their perception of whether that technology will be respectful.

A large body of research shows that user's technology-adoption behavior is often misaligned with their expressed privacy concerns. While this phenomena, the privacy paradox, is explained in part by the effect of many cognitive biases including endowment and ordering effects (Acquisti et al. 2013), it should perhaps not be such a surprise that people's decision to adopt or reject a technology is based on more than just the privacy of that technology.

Privacy calculus theory (PCT) agrees, going beyond considering just privacy to also consider benefits, arguing that "individuals make choices in which they surrender a certain degree of privacy in exchange for outcomes that are perceived to be worth the risk of information disclosure" (Dinev and Hart 2006). However, as I illustrate below, placing privacy as the sole detractor from adopting a technology and outcomes (or benefits) on the other remains too reductionist to fully capture user behavior, especially in well-being-related settings.

The incompleteness of a privacy-only view toward designing respectful technologies was exemplified in the rush to create COVID-19 technologies. In late 2020 and early 2021, technology companies and researchers developed exposure notification applications that were designed to detect exposures to coronavirus and notify app users of these exposures. These apps were created to replace and/or augment manual contact tracing, which requires people to call those who have been exposed to trace their networks of contact.

In tandem with the push to design these technologies was a push to ensure that these designs were privacy preserving (Troncoso et al. 2020). While ensuring the privacy of these technologies was critically important for preventing government misuse and human rights violations, and addressing user's concerns, people rarely adopt technologies just because they are private (Abu-Salma et al. 2017). Indeed, after many of these apps were released, a minority of people adopted them. Missing from the conversation was a discussion of user's other expectations for COVID-19 apps.

Privacy calculus theory posits that users trade off privacy against benefits and, in so doing, make decisions about what technologies to adopt. However, empirical research on people's adoption considerations for COVID-19 apps finds a more complex story (Li et al. 2020; Redmiles 2020; Simko et al. 2020). People consider not only the benefits of COVID-19 apps—whether the app can notify them of a COVID exposure, for example—but also how the efficacy of the app, how many exposures it can detect, might erode those benefits. Indeed, preliminary research shows that efficacy considerations may be far more important in user's COVID-19 app adoption decisions than benefits considerations (*Learning from the People: Responsibly Encouraging Adoption of Contact Tracing Apps* 2020). On the other hand, privacy considerations are not the only potential detractors; people also consider costs of using the system both monetary (e.g., cost of mobile data used by the app) and usability-related (e.g., erosion of phone battery life from using the app).

People's adoption considerations for COVID-19 apps exemplify the idea of respectful technologies: those that provide a benefit with a sufficient level of guarantee (efficacy) in exchange for using the user's data—with the potential privacy risks resulting from such use—at an appropriate monetary and usability cost. While COVID-19 apps offered benefits and protected user privacy, app developers and jurisdictions initially failed to evaluate the efficacy and cost of what they had built and failed to be transparent to users about both the efficacy and costs of these apps. As a result, people were unable to evaluate whether these technologies were respectful and the adoption rate of a technology that had the potential to significantly benefit individual and societal well-being during a global pandemic remained low.

Examining the full spectrum of people's respectful technology-related considerations is especially critical for well-being-related applications for two reasons.

First, there are a multitude of types of well-being that are increasingly being addressed by technology—from natural disaster check-in solutions through mental health treatment systems—each with a corresponding variety of different harms, costs, and risks that users may consider. If we focus strictly on the privacy-benefit tradeoffs of such technologies, we may miss critical adoption considerations such as whether the user suspects they might be harassed while using, or for using, a particular technology (Redmiles et al. 2019). Failing to design for and examine these additional adoption considerations can be a significant barrier to increasing adoption of commercially profitable and individually, or societally, beneficial technologies.

Second, different aspects of respectful technologies are prioritized by different sociodemographic groups (*Learning from the People: Responsibly Encouraging Adoption of Contact Tracing Apps* 2020). For example, older adults focus more on the costs of COVID-19 apps than do younger adults; younger adults focus more on the efficacy of these apps than do older adults. Ignoring considerations aside from privacy, and benefits, can perpetuate inequities in whose needs are designed for in well-being technologies and, ultimately, who adopts those technologies. Such equity considerations are especially important for well-being technologies for which equitable access is critical and for which inequitable distribution of harms can be especially damaging.

Thus, to ensure commercial viability and adoption of well-being technologies, and to avoid perpetuating and magnifying well-being inequities through the creation of such technologies, it is critical to build respectful well-being technologies. Technology creators and researchers must not only consider the privacy risks and protections of such technologies—and the technology's benefits—but also the contextual, cost, and efficacy considerations that together make up a potential user's view of whether a well-being technology is respectful of them and their data. To do so, two approaches are necessary: first, direct measurement of the cost and efficacy of technologies produced, in line with expectations for evidence from other fields such as health (Burns et al. 2011), and second, direct inquiry with potential users to understand contextual and qualitative costs. By combining these two approaches to empirical measurement, we can better create well-being technologies that are both effective and respectful.

References

Abu-Salma, R., Sasse, M. A., Bonneau, J., Danilova, A., Naiakshina, A., and Smith, M. (2017). Obstacles to the Adoption of Secure Communication Tools. *In*: *Security and Privacy (SP), 2017 IEEE Symposium on (SP17)*. *IEEE Computer Society*.

Acquisti, A., John, L. K., and Loewenstein, G. (2013). What Is Privacy Worth? *The Journal of Legal Studies*, 42 (2), 249–274.

Burns, P. B., Rohrich, R. J., and Chung, K. C. (2011). The levels of evidence and their role in evidence-based medicine. *Plastic and reconstructive surgery*, 128 (1), 305.

Byambasuren, O., Sanders, S., Beller, E., and Glasziou, P. (2018). Prescribable mHealth apps identified from an overview of systematic reviews. *npj Digital Medicine*, 1 (1), 1–12.

Dinev, T. and Hart, P. (2006). An Extended Privacy Calculus Model for E-Commerce Transactions. *Information Systems Research*, 17 (1), 61–80.

Learning from the People: Responsibly Encouraging Adoption of Contact Tracing Apps (2020). Available from: https://www.youtube.com/watch?v=my_Sm7C_Jt4&t=366s [Accessed 17 Mar 2021].

Li, T., Cobb, C., Jackie, Yang, Baviskar, S., Agarwal, Y., Li, B., Bauer, L., and Hong, J. I. (2020). What Makes People Install a COVID-19 Contact-Tracing App? Understanding the Influence of App Design and Individual Difference on Contact-Tracing App Adoption Intention. *arXiv:2012.12415 [cs]* [online]. Available from: http://arxiv.org/abs/2012.12415 [Accessed 17 Mar 2021].

Redmiles, E. M. (2020). User Concerns 8 Tradeoffs in Technology-facilitated COVID-19 Response. *Digital Government: Research and Practice*, 2 (1), 6:1–6:12.

Redmiles, E. M., Bodford, J., and Blackwell, L. (2019). "I Just Want to Feel Safe": A Diary Study of Safety Perceptions on Social Media. *Proceedings of the International AAAI Conference on Web and Social Media*, 13, 405–416.

Simko, L., Chang, J. L., Jiang, M., Calo, R., Roesner, F., and Kohno, T. (2020). COVID-19 Contact Tracing and Privacy: A Longitudinal Study of Public Opinion. *arXiv:2012.01553 [cs]* [online]. Available from: http://arxiv.org/abs/2012.01553 [Accessed 17 Mar 2021].

Troncoso, C., Payer, M., Hubaux, J.-P., Salathé, M., Larus, J., Bugnion, E., Lueks, W., Stadler, T., Pyrgelis, A., Antonioli, D., Barman, L., Chatel, S., Paterson, K., Čapkun, S., Basin, D., Beutel, J., Jackson, D., Roeschlin, M., Leu, P., Preneel, B., Smart, N., Abidin, A., Gürses, S., Veale, M., Cremers, C., Backes, M., Tippenhauer, N. O., Binns, R., Cattuto, C., Barrat, A., Fiore, D., Barbosa, M., Oliveira, R., and Pereira, J. (2020). Decentralized Privacy-Preserving Proximity Tracing. *arXiv:2005.12273 [cs]* [online]. Available from: http://arxiv.org/abs/2005.12273 [Accessed 17 Mar 2021].

Vitak, J., Liao, Y., Kumar, P., Zimmer, M., and Kritikos, K. (2018). Privacy attitudes and data valuation among fitness tracker users. *In: International Conference on Information.* Springer, 229–239.

Part XI
Realizing Digital Humanism

Digital Humanism: Navigating the Tensions Ahead

Helga Nowotny

Abstract The assumption of digital humanism that a human-centered approach is possible in the design, use, and further development of AI entails an alignment with human values. If the more ambitious goal of building a good digital society along the co-evolutionary path between humans and the digital machines invented by them is to be reached, inherent tensions need to be confronted. Some of them are the result of already existing inequalities and divergent economic, social, and political interests, exacerbated by the impact of digital technologies. Others arise from the question what makes us human and how our interaction with digital machines changes our identity and relations to each other. If digital humanism is to succeed, a widely shared set of practices and attitudes is needed that sensitize us to the diversity of social contexts in which digital technologies are deployed and how to deal with complex, non-linear systems.

The availability of masses of data, efficient algorithms, and unprecedented computational power has pushed humans on a co-evolutionary path with the digital machines we have created. Seen from an evolutionary perspective, this might look like another of the many evolutionary trials and errors whose outcome leads either toward a dead end or toward new forms of life. Although this is impossible to predict, we should remind ourselves that cultural evolution, spearheaded by science and technology, has overtaken biological evolution. It has equipped the human species with cognitive capabilities that have enabled it to generate the digital entities, devices, and infrastructures with which humans interact in ever-more intricate and intimate ways. We should know them better than they know us—yet, we are repeatedly plagued by the anxiety that in the end they might dominate us.

Thus, we oscillate between trust in the digital technologies that have become our daily companions while being aware that there are many reasons for distrust and caution. Concerns about privacy and fear of surveillance co-exist with the collusion

H. Nowotny (✉)
ETH Zürich, Zurich, Switzerland
e-mail: helga.nowotny@wwtf.at

© The Author(s) 2022
H. Werthner et al. (eds.), *Perspectives on Digital Humanism*,
https://doi.org/10.1007/978-3-030-86144-5_43

of voluntarily handing over our data to the large corporations (Zuboff 2018). The possibilities of abuse and malfunctioning of vulnerabilities to hacking and other forms of cyber-insecurity persist, while optimistic scenarios of new opportunities continue to be acclaimed. We rightly insist that in critical situations, humans ought to be the ones whose judgement trumps automated response and decisions and that accountability must be built into the process in case things turn bad (Christian 2020; Russell 2019). On this co-evolutionary journey and despite the uncertainty of its outcome, we feel encouraged by what may turn out to be an illusion: that we have been dealt the slightly better cards in the co-evolutionary game and human ingenuity will prevail. This is one of the premises on which digital humanism rests, the belief that human values can be instilled in the digital technologies and that a human-centered approach will guide their design, use, and future development (Werthner et al. 2019).

For digital humanisms, such aspirations serve as the necessary preconditions for gaining momentum, but ought not obscure the difficulties that lie ahead. In the long history of technological inventions and innovations, humans always attempted to be in control. What began as deploying tools thousands of years ago to carve out a precarious living from the natural environment turned into massive intervention and large-scale change of the natural environment during industrialization, with devastating consequences for the latter on which we depend. The peak of the belief that humans were in complete control of technology and mastering their future came during modernity (Scott 1999). A turning point was reached in the mid-twentieth century, when it became clear that we were no longer in control over the radioactive waste left behind from the production of the atomic bomb. After the end of the war, the world's population began to grow dramatically, and so did GDP and living standards. At the same time, the impact of human intervention in the earth system and its services began to be noticeably felt. Called "The Great Acceleration," the convergence of these two large-scale developments has not abated since (Steffen et al. 2015; McNeill and Engelke 2015). Today, we are faced with a major sustainability crisis, while digitalization is rapidly gathering momentum with profound and far-reaching implications for what it means to be human and what a good digital society could be. We have arrived in the Anthropocene, and it will be a digital Anthropocene.

Digital humanism thus emerges at a crucial moment, at the intersection of the sustainability crisis and the opportunities offered by digitalization. In order to gauge the challenges it faces, we ought to remind ourselves of the continuities and ruptures it entails. It aspires to build upon some of the great cultural transformations that are part of the European heritage, exploring human nature and adopting a human-centered approach under rapidly changing global circumstances. But digital humanism also harbors a rupture that is less obvious. It marks the transition from the linearity in thinking and understanding the world which was one of the hallmarks of modernity toward coping with the non-linear processes of complex adaptive systems. Just as it is no longer possible to rely on the linearity of technological progress that will inevitably lead to a future being better than the past and the present, digital humanism must guide us in thinking in non-linear terms when we increasingly face uncertainty and complexity (Nowotny 2015).

Digital humanism therefore must navigate the different strands of our existence that emerge from the inherent tension between humans and machines. In philosophical terms, we speak about life and non-life, about organic and inorganic matter, about different rates of energy conversion needed to keep us and machines going, and, ultimately, about consciousness and their absence in machines (Lee 2020). But as there is little agreement on the definition of these terms and their meaning, the entangled interaction between humans and the digital machines continues in practice as a blurred and messy process. Digital humanism, if it is to be enacted, must be prepared to navigate the manifest and hidden tensions that come to the fore in expected and unexpected ways and in different constellations.

Digitalization exacerbates already existing and familiar tensions between divergent economic, political, and social interests, as amply demonstrated during the COVID-19 pandemic when societal inequalities and fissures were laid bare. Fake news and conspiracy theories continue to circulate freely in the social media, turning science into mere opinion and risking the further destabilization of already fragile liberal democracies. Many unresolved conflicts are linked to rising inequalities. As the digital divide deepens, the fear persists that digitalization will replace jobs faster than new ones will be generated (Susskind 2020). These manifest tensions can ignite major conflicts and further tear apart the social fabric already under considerable stress. Digital humanism cannot abstain from entering this contested arena. It cannot retreat to pursue the ideal of a humanistic and digitally sophisticated individual without considering the digital society that shapes how we live together. Digital humanism will have to come up with designs for new modes of digital governance that can meet the challenges of what a good digital society fit for the twenty-first century could be.

Other tensions are less visible; some are latent or emerging. They hover above the question that constitutes the core of digital humanism: what makes us human and how does the interaction with digital machines change us? Some of these tensions fuel identity anxieties that are directly or indirectly related to social media or the feeling that an algorithm knows us better than we know ourselves. If the experience of acceleration dominated modernity, the prevailing experience in the digital age is informational overload and emotional overextension. As the past reaches into the present and the future has already arrived, at least in the visible form of the latest digital devices, the present becomes densified and further compressed. Digital humanism is challenged to create new spaces in this overheated and hyper-reactive atmosphere in which physical presence needs to be reconciled with virtual space in ways that have yet to be invented. The virus has taught us much about the needs of our bodies in a digital world. Whatever lessons we draw, digital humanism will have to open new pathways for implementing them.

The uncanny efficiency of predictive algorithms and their practical take-over in decision-making pervading our individual and collective life mark another tension-ridden domain for digital humanism to navigate. Whether it is the entire health sector or individual lifestyles, our consumption behavior, or the functioning of our institutions, predictive algorithms extrapolate from the past to let us see further ahead into the future. Yet, in doing so, they lure us into transferring agency unto them. Once we start to believe that an algorithm can predict what will happen in the future and supportive digital decision-making systems are widely adopted, the point may be reached when human judgement seems superfluous and algorithmic predictions turn into self-fulfilling prophecies (Nowotny 2021).

Thus, the stakes for a digital humanism are high. In order to navigate these tensions, it will have to come up with concrete propositions that include the deeper humanistic layers, going beyond technological solutions. Important as appeals to ethical principles are, they will not suffice unless they can draw in very practical terms on a widely shared set of attitudes and practices that are inspired and guided by a humanistic ideal as a way of living together. This involves devising new ways of tackling problems that go beyond technological fixes and to acknowledge that "wicked problems" exist for which no solutions are in sight, yet they too must be confronted. Digital humanism draws its strength from the conviction that a better digital society is possible, mustering the courage to experiment with what is needed for shaping it.

In practice, this means to cultivate a humanistic sensitivity for the diversity of social contexts in which digital technologies are deployed and efficacious. Currently, neither predictive algorithms nor the data on which they train are sufficiently context-sensitive. Digital humanism can let us discover hitherto unknown features of who we are without determining what we will be. It can teach us the irreplaceable value of critical human judgement when we face the illusionary promise of predictive algorithms that they know the future, which is not determined by any technology but remains uncertain and open. The major benefits of digital processes do not only consist in being "smart," but other potential benefits wait to be explored with an open and curious mind. Digital humanism can sensitize us how to deal with complexity which is closer to our intuitive understanding of what is means to be human than a linear, cause-effect way of thinking. It can attune us to emergent properties and to what remains unpredictable—the ultimate sign of life that keeps evolving.

References

Christian, B. (2020) The Alignment Problem. Machine Learning and Human Values. New York: Norton & Company.

Lee, E. A. (2020) The Coevolution: The Entwined Futures of Humans and Machines. Cambridge, MA: MIT Press.

McNeill, J. R. and Engelke, P. (2015) The Great Acceleration: An Environmental History of the Anthropocene since 1945. Cambridge, MA: Harvard University Press.

Nowotny, H. (2015) The Cunning of Uncertainty. Cambridge, UK: Polity Press.

Nowotny, H. (2021) In AI We Trust. Power, Illusion and Control of Predictive Algorithms. Cambridge, UK: Polity Press.

Russell, S. (2019) Human Compatible: AI and the Problem of Control. London: Allen Lane.

Scott, J 1999, Seeing Like a State: How Certain Schemes to Improve the Human Condition Have Failed, Yale University Press, New Haven.

Steffen, W. et al. (2015) 'The Trajectory of the Anthropocene: The Great Acceleration', The Anthropocene Review 2:1, pp. 81–98.

Susskind, D. (2020) A World Without Work: Technology, Automation, and How We Should Respond. London: Allen Lane.

Werthner, H et al. (2019) Vienna Manifesto on Digital Humanism, viewed 20 June 2019, <www.informatik.tuwien.ac.at/dighum/>.

Zuboff, S. (2018) The Age of Surveillance Capitalism: The Fight for a Human Future at the New Frontier of Power, New York: Hachette Book Group.

Should We Rethink How We Do Research?

Carlo Ghezzi

Abstract Advances in digital technologies move incredibly fast from the research stage to practical use, and they generate radical changes in the world, affecting humans in all aspects of their life. The chapter illustrates how this can have profound implications on the way technological research is developed. It also discusses the need for researchers to engage more actively in public debates with society.

1 Introduction: Coping with Disruptive Changes

History shows that advances in science and technology have always produced changes in the world in which we live. With digital technologies,[1] we experience unprecedented levels of change. They can affect human well-being, assist humans in their individual activities, and relieve them of strenuous or dangerous jobs. More broadly, they create an entirely new *cyber-physical world* in which they live and interact with natural phenomena, other individuals, and new kinds of artificial autonomous entities. The "old world," which was known to us for centuries, in which we felt comfortable and with which we slowly evolved, has been suddenly replaced by a new one whose laws we ignore and where humans may lose control. Digital humanism stresses that humankind must be at the center of innovation and technology has to serve society and the environment.

Technological developments are largely driven by research. To understand the lessons for research we can learn from the development of digital technologies, we need to reflect on its two key properties: radicality and speed of change.

The effects of digital technologies on humankind have indeed been revolutionary. Other radical shifts were generated by research in the past, but perhaps none is

---

[1]The umbrella term "digital technologies" includes both hardware and software (data, algorithms, AI).

C. Ghezzi (✉)
Politecnico di Milano, Milano, Italy
e-mail: carlo.ghezzi@polimi.it

© The Author(s) 2022
H. Werthner et al. (eds.), *Perspectives on Digital Humanism*,
https://doi.org/10.1007/978-3-030-86144-5_44

affecting as profoundly every human life. For example, the shift from Ptolemaic astronomy—which considered the earth to be at the center of the universe with the sun, moon, and planets revolving around it—to the Copernican view had radical effects on science and philosophy but did not affect the individual's everyday life. Likewise, the radical shift in physics at the beginning of the twentieth century, caused by the developments of relativity theory and quantum mechanics, challenged the known view of the physical world, described by the axioms and laws of Newton's mechanics. This was a spectacular paradigm shift, which however had little effect on practically observable phenomena in everyday human life.

In addition, the transition of advances in digital technologies from the research stage into everyday life occurred much faster than for previous technologies. As observed, for example by Harari (2014), the Industrial Revolution, ignited by the invention of steam engines, took about two centuries to develop the modern industrial world. Digital technologies spread to the world in only a few decades, generating abrupt changes and hampering gradual and friction-less adaptation.[2]

2 Effects on How We Do Research

The main implication of speed of change is that scientists and engineers,[3] in their exploratory work, cannot ignore the potential implications and effects of the new technology they develop. Delaying reflection on the use of technology to a later stage may cause serious harm. Alas, "later" is in fact "sooner" than expected; it can be too late! Rather, technological research must proceed hand-in-hand with the investigation of its implications. Traditionally, careful deployment strategies and trial usage of newly developed technology could prevent major damage, through adjustments and countermeasures. Today, however, radical innovations in image recognition based on AI deep learning techniques have been immediately transferred to practice without first exploring their limitations and potential implications. Adoption in law trials has raised serious ethical concerns and potential violations of human rights. Likewise, the advances that enabled mass mobile pervasive computing through personal devices and smartphones stressed mainly usability and functionality, at the expense of trustworthiness and dependability. As a result, infrastructures are open to all kinds of misuses and attacks, including privacy violations that led to serious political consequences.[4] Plenty of examples of ethically sensitive technical issues are faced by current research on automatic vehicles.

---

[2]The digital revolution is also very fast and radically changing the way we do research, in almost all areas, through the unprecedented availability of data and the invention of algorithms which can manipulate them and reason about them, leading to discovery automation. The deep consequences of this change would require further discussion.

[3]In this chapter, the terms scientist and researcher are used interchangeably. Furthermore, they mainly refer implicitly to technological research.

[4]For example, the Pegasus spyware [https://en.wikipedia.org/wiki/Pegasus_(spyware)].

The effect on humans and society of radical changes is a serious concern that must be addressed while developing technological research. To deal with it, research has to broaden its focus, moving beyond pure technical boundaries and bringing in a focus on the potential human and societal issues involved. This asks scientists to break the rigid silos into which research is currently compartmented. Philosophers and social scientists, for example, need to be integrated in research groups that develop new technology for autonomous vehicles; environmentalists, urban planners, and social scientists need to work with computer scientists to develop traffic management solutions in smart cities. The quest for interdisciplinarity—so far often more a fashionable slogan than reality—becomes a necessity. We seriously and urgently need to understand how this can be done. For example, how to achieve breadth without sacrificing depth of research, how to evaluate interdisciplinary work without penalizing it by applying traditional silo-based criteria, etc.

3 Effects on Engagement with Society

Speed and radicality of change have an important consequence on the need for scientists to engage with society. Traditionally, they interact and communicate almost exclusively with their silo peers. They are largely shielded from direct communication and interaction with a broader public. Limited forms of engagement include innovation initiatives, like generation of spin-offs and collaborations with industry, government, and policy makers. Fast and radical changes require more involvement, especially in discussing the potential developments and uses and raising broad social awareness. This, however, is easier said than done.

Researchers know well how to communicate with their peers. They learn how to do it since they enter a PhD program and continue to learn and improve throughout their career. Research is an intrinsically open process that relies on communication among peers. The main ambition of scientists is to achieve novel results and communicate them to the research community trough research papers, artifacts (such as data sets or software prototypes), and scientific debates in conferences. Their career progress largely depends on how successful they are in producing and spreading novel and relevant results to their peers. Ghezzi (2020) discusses the importance of communication among peers and also stresses the need for more neglected forms of public engagement, through which scientists lead, or participate in, scientific debates with a broader audience, outside the circle of peers: with government, policy and decision-makers, and the general public.

There are notable historical examples of scientists who engaged in public scientific debates, especially when progress led to radical changes. A famous case is Galileo Galilei, who strove to bring his support to the Copernican theory to the attention of the society of his time. He was well aware of the profound consequences of the shift from the Ptolemaic to the Copernican view: mankind was no more living at the center of the universe, but instead on a planet, which was just a small part of the solar system. He spoke to the informed society of his time through an essay,

"Dialogue Concerning the Two Chief World Systems," in which a scientific conversation is carried on among three individuals: a Copernican scientist who explains the new theory to an educated citizen arguing against the statements made by a Ptolemaic scientist. Galileo is considered as the father of modern science. He taught to us that science is not blind faith on previous beliefs, dominant orthodoxy, or ideology. It relies on rational argumentations to arrive at whatever conclusions a careful analysis of evidence would suggest, even if they are not conformant with current beliefs. He developed new technology to empower humans to understand and dominate the physical world. His public engagement led him to confront the Catholic official doctrine, which followed the Ptolemaic view that the earth was at the center of the solar system. Galileo appeared before the Roman Inquisition and was eventually accused of heresy. He was forced to recant his views and sentenced to house arrest for the rest of his life.

Another example of a heated public debate occurred at the beginning of the twentieth century when the revolutionary developments in physics and mathematics by giants like Einstein, Plank, Bohr, and Hilbert brought together an outstanding group of physicists, philosophers, and mathematicians, who met in Vienna in a permanent seminar from 1924 until 1936, called Wiener Kreis (Vienna Circle). The members of the seminar aimed at founding philosophy on a modern scientific view of the world, keeping it separate from metaphysics.

Participants in the discussions included physicists, philosophers like Schlick (who chaired the group), Neurath, Popper, and Wittgenstein and mathematicians and logicians like Carnap and Gödel. The Vienna Circle dissolved in 1936, when Schlick died, and anti-semitism caused a diaspora of the other members. A fascinating account of this highly influential movement can be found in Sigmund (2017). More heated public discussions involved physicists at the end of World War II, when the relation between research and its direct use in the development of mass destruction weapons became evident. The debate was able to inform and involve both governments and citizens.

These examples of public involvement in scientific debates remained mostly at the level of "educated elites." The digital revolution is directly affecting every individual's life and requires a broader reach out. An informed debate needs to take place, involving not only scientists in almost all areas and decision-makers at all levels, but also citizens, to make sure that humans and the earth on which they live are at the center of technological developments.

4   Conclusions

Effective engagement with the general public requires that researchers learn how to communicate effectively and that their efforts in doing so are recognized and rewarded. They need to understand the role they must play in this conversation, which mainly aims at explaining the advances of research and pointing to the critical issues involved in its use, which may require collective, informed, rational decisions.

The boundaries between scientific knowledge and personal opinions and beliefs should be kept clearly separate. Effective communication also demands a mature and competent audience. This raises serious concerns, since regrettably the level of scientific education has been decreasing in many countries. Even worse, paradoxically, in our highly technological world there is a widespread mistrust in science—see Nichols (2017)—which should be contrasted by more investments in education. In particular, we need to ensure that every responsible citizen understands how science and technology progress, how they can be trusted, and what their limits are. Development of an open space for discussions around digital technologies is crucial for the future of our democratic societies and realization of digital humanism.

References

Ghezzi, C. (2020) *Being a Researcher: An Informatics Perspective.* Springer International Publishing.

Harari, Y.N. (2014) *Sapiens: A Brief History of Humankind.* Random House.

Nichols, T. (2017) *The Death of Expertise: The Campaign against Established Knowledge and Why it Matters.* Oxford University Press.

Sigmund, K. (2017) *Exact Thinking in Demented Times: The Vienna Circle and the Epic Quest for the Foundations of Science.* Basic Books.

Interdisciplinarity: Models and Values for Digital Humanism

Sally Wyatt

Abstract This chapter starts from the recognition that the world is facing major challenges and that these may be best addressed by people working together, across different disciplinary domains and between universities, civil society, governments, and industry. After sketching these problems, I provide an overview of the meanings of discipline and of multi-, inter-, and transdisciplinarity. I then provide a brief historical overview of how disciplines emerge. Examples from computer sciences, social sciences, and the humanities, and collaborations between them, are used to illustrate these definitions and overview. In the final part, I reflect on what this means for digital humanism, drawing on different models and values of collaboration.

The 2030 Agenda for Sustainable Development, prepared by the United Nations (UN) and approved by all countries in 2015, identifies 17 goals, crucial for the future of the planet. These include ending poverty, empowering women and girls, reducing inequality, and taking action to combat climate change (UN 2015). Interestingly, digital technologies are not explicitly mentioned in any of the goals, although they could be seen as both part of the problem given, for example, their enormous energy needs and the emergence of new forms of digital inequality. They could also be part of the solution by making it easier to share data and knowledge to solve problems such as those arising from an ageing population and by expanding access to education.

The problems underlying these goals could be characterized as "wicked problems," those political and intellectual challenges that defy easy definition or solution. No single academic discipline can provide an adequate definition of such problems much less a clear and feasible resolution. The UN calls for partnership and collaboration to tackle these goals. To do so will require multi-, inter-, and transdisciplinary research. The UN is not alone in making such calls. Many research-funding agencies and policy-making organizations emphasize the importance of engaging

S. Wyatt (✉)
Maastricht University, Maastricht, The Netherlands
e-mail: sally.wyatt@maastrichtuniversity.nl

with different disciplines and stakeholders in order to tackle contemporary social and scientific problems.

In this short chapter, I first discuss the meaning of discipline and of multi-, inter-, and transdisciplinarity. I then provide a brief historical overview of how disciplines emerge and conclude with different models and values of collaboration and what these could mean for digital humanism.

Multi-, inter-, and transdisciplinarity are sometimes used interchangeably, but they each capture something different, described in the following paragraphs. But first it is necessary to understand what an academic discipline is. Disciplines have their own cultures and practices and provide those trained in them with skills, tools, methods, concepts, and ways of thinking. They come with their own notions of how the world is organized and of what constitutes good quality research (Knorr Cetina 1999). Disciplines are usually institutionalized, in university departments and faculties, in professional associations, and in specialized conferences and journals. Reproduction of disciplines from one generation to the next is typically done via formal, accredited education and sometimes involves shared competence criteria (Hackett et al. 2017). An example of the latter is the "Computing Competencies for Undergraduate Data Science Curricula" (ACM 2021). It is less usual to find such criteria in the humanities and the social sciences, although those disciplines often have implicit norms and expectations of the knowledge and competences students should possess by the end of their degree programs.

Having provided a working definition of discipline, let us now move to the ways they may be combined. These are usually presented in a hierarchical form, with multidisciplinarity being the least integrated. Multidisciplinarity can be described as moving between disciplines in order to understand a topic or problem from different perspectives. This can lead to greater knowledge and may be very helpful in making policy or other decisions, but there is little integration of methods or concepts from the contributing disciplines. For example, economic modelling can be used to understand the incidence of poverty in a country, but pedagogical studies provide the basis for policies to tackle educational inequalities between children from different socioeconomic backgrounds.

Interdisciplinary education and research deliberately attempt to combine and synthesize methodologies and specialized jargon from different disciplines in order to produce a more comprehensive solution to a problem or to address a complex topic. For example, this occurs when computer scientists and linguists work together to understand changing language patterns in large text corpora.

Transdisciplinarity goes outside the university in order to incorporate knowledge from other non-academic sources and stakeholders. There are many possible stakeholders with specialized knowledge and experiences that can be valuable in the production of knowledge. In the case of healthcare, this could include patient organizations, the pharmaceutical industry, and nursing professional associations or unions as well as biomedical researchers, sociologists of health, medical ethicists, and data scientists. There are many terms in circulation for transdisciplinary knowledge production, including post-normal science (Funtowicz and Ravetz 1993), the triple helix (Leydesdorff and Etzkowitz 1998), and Mode 2 (Nowotny et al. 2001).

Having briefly defined the key terms, let us return to academic disciplines. They can have the appearance of immutability, rather like an immutable object in some kinds of computer programming, something that cannot be changed after it has been created. Nonetheless, it is important to remember that academic disciplines can and do change. Many academic disciplines now taken for granted, such as mathematics, history, and philosophy, have very long histories, just as universities do. Others, including engineering and social sciences, emerged in the late nineteenth century, largely in response to the challenges posed by industrialization and urbanization in Europe and the United States. The rise of industrialization and engineering was in no small part the impetus behind the establishment of technical universities in many countries. Even though change might be slow, new disciplines can and do emerge, and the focus and emphasis in long-standing disciplines may change.

The expansion of the university system after World War II in industrialized countries was one catalyst for change. Growth was accompanied by an increase in the diversity of students, staff, and (inter)disciplines. In the final third of the twentieth century, the emergence of a new field was sometimes related to the diffusion of a new object, such as the internet in the case of new media studies. In other cases, the availability of new techniques and instruments could lead to a new field, as in computer science. In yet other cases, such as women's studies, the emergence could be attributed to the greater diversity of people entering universities, people who may identify new problems and ways of working (Wyatt et al. 2013). Such new fields may find their first institutional homes in literature, electrical engineering, or sociology. As they grow and stabilize, they can become institutionalized in the ways mentioned above, by developing their own departments, educational programs, journals, and professional associations.

Not all attempts at creating new disciplines are successful. Some might be very strong in research and the creation of new knowledge, published and shared in specialized journals and conferences, but this is not always accompanied by widespread or strong educational profiles. For example, neuroeconomics—the study of how economic behavior affects understanding of the brain and how neuroscience might guide economic models—might be taught only in a relatively small number of universities at advanced level. Nonetheless it has its own specialized jargon, with conference and publication outlets, for sharing ideas and developments.

Inter- and transdisciplinary collaborations have, as mentioned above, been heralded by national and international organizations looking for innovative solutions to complex and wicked problems. But collaborations are not always easy to achieve. Not all disciplines are equal, neither in terms of available funding nor in terms of epistemic and social legitimacy and status. Such inequalities can hinder productive collaboration, and thus the remainder of this text focuses on different modes of collaboration across disciplines.

In particular, I reflect on what this might mean for "digital humanism." that "community of scholars, policy makers, and industrial players who are focused on ensuring that technology development remain centred on human interests."[1] According to the definitions sketched above, this is clearly a transdisciplinary endeavor, retaining the problem-solving aspirations of engineers and computer scientists, but doing so in a way that supports human interests and well-being. These are, in part, captured by the UN's Sustainable Development Goals mentioned earlier. They could also be articulated in terms of fundamental values of democracy, equality, freedom, and solidarity.

Just as there are different ways of doing disciplinary research (and it must be remembered that disciplines are not homogenous in their methods and theories), there are different ways of doing inter-, multi-, and transdisciplinarity. Barry et al. (2008) distinguish between three modes of interdisciplinarity: service-subordination, integration-synthesis, and agonistic-antagonistic. In the service-subordination mode, one discipline contributes to another without changing the rules of the discipline to which it is contributing. For example, computational methods and tools could be taken up within linguistics without any fundamental change to linguistic theory. Or, a historian could explain the boundary changes between cities or regions or the development of occupational categories that make the merging of historical census data more difficult.

The integration-synthesis mode refers to a more symmetrical relationship between the contributing disciplines through a genuine integration of methods and concepts, as in the case of neuroeconomics mentioned above, and in countless other examples. For example, in order to understand how researchers find data for potential re-use, Gregory (2021) draws on both information science and science and technology studies (STS) to develop a richer understanding of the diversity of users and their data practices.

In the third antagonistic mode, those from one discipline might aim to alter another in fundamental ways. This has sometimes been claimed in efforts to bring computational ways of thinking to the humanities, for example. At its worst, this can be seen as insulting, by suggesting that computational notions of rigor and reliability are superior to the quality standards of the humanities. But antagonism does not have to be negative. Academic research, in all disciplines, is characterized by debate and by careful scrutiny of knowledge claims and of the evidence on which they are made. It is to be expected that digital humanism will be characterized by sometimes heated debates between computer scientists wishing to solve what they define as technical problems and those in the humanities and social sciences who might find this excessively narrow and will point out how entangled the social and technical always are. In other words, the social scientist or humanities scholar might argue that digital technologies are always a material intervention in society and cannot be understood independently from social, cultural, economic, and political contexts. This can be productive as it can lead to reconfiguring the "boundaries, objects, and problematics" of the antecedent disciplines (Barry et al. 2008, p. 30).

---

[1] https://dighum.ec.tuwien.ac.at/

These different modes of interdisciplinarity are intended as heuristic. They are not exhaustive of modes of collaboration and nor are they mutually exclusive. From my own experiences of interdisciplinary collaboration, I have identified one resource and two values: time, respect and humility. There are many guidelines regarding collaboration, but there is no fast or simple route to success. Just as disciplinary training takes time, so too does learning to collaborate. Each project or group needs time to develop shared vocabularies, methods, and ways of working. People also need to respect other disciplinary ways of working even if they do not necessarily understand them. We all need to recognize that the disciplines in which we have been trained may not have all the answers nor even always the right questions. This is another way of phrasing the old adage that "if your only tool is a hammer, then everything looks like a nail."

References

ACM Data Science Task Force (2021). 'Computing competencies for undergraduate data science curricula', Available at: DSTF_Final_Draft_Report (acm.org) (Accessed: 21 March 2021)

Barry, A., Born, G. and Weszkalnys, G. (2008). 'Logics of interdisciplinarity', *Economy & Society*, 37(1), 20-49.

Funtowicz, S. and Ravetz, J. (1993). 'Science for the post-normal age', *Futures*, 25, 739-755.

Gregory, K. (2021). *Findable and reusable? Data discovery practices in research*. PhD thesis. Maastricht University.

Hackett, E.J. et al. (2017). 'The social and epistemic organization of scientific work', in Felt, U. et al. (eds.) *The handbook of science and technology studies, 4th edition*. Cambridge, MA: The MIT Press, pp. 733-764.

Knorr Cetina, K. (1999). *Epistemic cultures: How the sciences make knowledge*. Cambridge, MA: Harvard University Press.

Leydesdorff, L. and Etzkowitz, H. (1998). 'The triple helix as a model for innovation studies', *Science & Public Policy*, 25(3), 195-203.

Nowotny, H., Scott, P. and Gibbons, M. (2001). *Re-thinking science. Knowledge and the public in an age of uncertainty*. London: Polity Press.

United Nations (2015). *Transforming our world: The 2030 agenda for sustainable development*. New York, NY: United Nations, Department of Economic and Social Affairs.

Wyatt, S. et al. (2013). 'Introduction to Virtual Knowledge', in Wouters, P. et al. (eds.) *Virtual knowledge. Experimenting in the humanities and the social sciences*. Cambridge, MA: The MIT Press, pp. 1-23.

It Is Simple, It Is Complicated

Julia Neidhardt, Hannes Werthner, and Stefan Woltran

Abstract History is not a strictly linear process; our progress as society is one full of contradictions. This we have to bear in mind when trying to find answers to pressing challenges related to and even caused by the digital transformation. In this chapter, we reflect on contradictory aspects of Digital Humanism, which is an approach to foster the control and the design of digital infrastructure in accordance with human values and needs. Seemingly simple solutions turn out to be highly complex when looking at them more closely. Focusing on some key aspects as (non-exhaustive) examples of the simple/complicated dilemma, we argue that, in the end, political answers are required.

History is a dialectic process. This is also true for the ongoing digital transformation (probably much better named informatization as it is informatizing nearly everything). Much of this development has already happened in the past, unobserved by mass media and most decision-makers in politics and industry. Nowadays, this transformation appears at the surface, leaving many with the impression of an automatism, of a process without human control, guided by some "external" forces. This is why the notion of Digital Humanism is so important. As an approach to counteract negative effects of the digital transformation, it aims to foster the control and the design of digital infrastructure in accordance with human values and needs. While numerous people support (or at least sympathize with) these general aims and goals of Digital Humanism, there are subtle questions when looking behind the scene that need to be discussed and resolved. We should resist the temptation to provide trivial solutions which will not prevail in a serious discussion.

Our progress as society is full of contradictions, and sometimes it even points backwards. Bearing in mind such contradictions and that historical processes do not move straight forward linearly, we try to approach some of these contradictory aspects of Digital Humanism by looking at it using the intertwined pair "simple"

J. Neidhardt (✉) · H. Werthner · S. Woltran
Vienna University of Technology, Vienna, Austria
e-mail: julia.neidhardt@tuwien.ac.at; Hannes.werthner@tuwien.ac.at;
stefan.woltran@tuwien.ac.at

© The Author(s) 2022
H. Werthner et al. (eds.), *Perspectives on Digital Humanism*,
https://doi.org/10.1007/978-3-030-86144-5_46

and "complicated." What seems to be simple is complicated, and vice versa. This chapter also reflects a discussion among us, the authors. Our sometimes-controversial debate may serve as a blueprint for a future "dialectic" process to find agreements on complicated matters, also in a public debate. The following, by no means exhaustive, list represents some of the "simple/complicated" issues Digital Humanism has to address:

- *Interdisciplinarity.* The impact of digitalization on our lives is obvious, and everybody senses its power (both positive and negative). Negative phenomena to be addressed include the monopolization of the Web, issues related to the automatization of work, problems with respect to AI and decision making, the emergence of filter bubbles, the spread of fake news, the loss of privacy, and the prevalence of digital surveillance. While these are all aspects of the same disruptive process, they manifest very differently. Consequently, a broad spectrum of challenges needs to be addressed. The obvious and simple conclusion is: Interdisciplinarity is needed to tackle them, to understand the complicated presence, and to shape the digital future.

  But is it really so simple, as interdisciplinarity brings its own challenges? It is very hard, for instance, to come up with a common language, where all researchers involved use the same terminology with the same meanings. The way, moreover, how the research landscape is organized, still hinders interdisciplinarity. Interdisciplinary (especially young) researchers often do not obtain funding since they touch different communities but are not specialized enough to be at their centers, which often leads to negative reviews; so how to foster interdisciplinarity for Digital Humanism, on a content, a method, and an institutional level? Even more, as Informatics—as a key discipline—often comes with the attitude to "solve problems," but at the same time not always seeing the side effects and long-term impacts of its work. Computer scientists cannot or even should not be the sole driving force. But if so, the role of Informatics and its methods needs some clarification. For a solid foundation of Digital Humanism, exchange across various disciplines is needed throughout the entire process, i.e., when doing analyses, when developing new technologies, and when adopting them in practice. Looking back in history, one sees that artifacts created by computer scientists have similar (if not even more, given their more pervasive nature) impact as the steam engine had in the Industrial Revolution. But it was not the engineers who organized the workers and envisioned social welfare measures; it was a much broader effort including intellectual leaders with diverse backgrounds together with the workers and their unions.

- *Humans decide.* As it is written in the manifesto, "Decisions with consequences that have the potential to affect individual or collective human rights must continue to be made by humans" (Werthner et al. 2019). In a world becoming more and more complicated and diverse, it is obvious that long-reaching and fundamental decisions should be made by humans. It is about us and our society—thus it is up to us, we are responsible for ourselves. This seems to be a simple principle.

However, it may be a little bit more complicated; empirical research in psychology and human decision-making shows (see Kahnemann 2011 or Meehl 1986) that rather simple statistical algorithms (e.g., multiple regression models) seem to outperform humans, even experts in the respective field, in particular when long-term decisions are to be taken. We humans tend to build causal models of the world, as a "simplification" for being able to understand the world. This is the case even or especially in complicated cases where randomness plays an important role. In addition, unobserved or unconsidered parameters can influence human decisions.[1] So the issue does not seem to be an either or nor, but rather how and when to combine humans and machines—complicated!

- *IT for the good.* Digital Humanism not only attempts to eliminate the "downsides of information and communication technologies, but to encourage human-centered innovation" (Werthner et al. 2019); its focus is also on making the world a better one to live in, to contribute to a better society. In essence, it is simple when taking for example the Corona crisis, where informatization has shown its positive potential (both in research as well as in enabling a further functioning of our society). And when one looks at the United Nations Sustainable Development Goals (SDG—https://sdgs.un.org/goals), one can see that they only can be achieved through proper IT research and its application.

But reality is again a little bit more complicated. Since the 1980s, income inequality has risen in practically all major advanced economies, and this is the period of the ongoing digitalization.[2] But not only has the gap within the society widened[3]—following the "rules" of the networked platform economy with its *winners take it all* principle—there is also a growing market gap between companies. For example, today the first seven most valued companies at the stock markets are IT platform companies; back in 2013 only two of them were in the top ten. In addition, the productivity growth has slowed down, where one would have expected substantial growth, as forecasted by several public relations companies in the IT field. Thus, also the cake of wealth to be distributed did not grow in order to "appease" the majority of the population.[4] It is difficult to isolate the causes of this socioeconomic development, but for sure technology plays a crucial role. So, it is complicated, in the analysis, and also in finding the proper technological and political answers.

---

[1] Danzinger et al. (2011) show that "judicial rulings can be swayed by extraneous variables that should have no bearing on legal decisions." When examining parole boards' decisions, the authors found that the likelihood of a favorable ruling is greater at the beginning of the work day or after a food break than later; so, whether before or after a break matter.

[2] See, e.g., www.bbvaopenmind.com/en/articles/inequality-in-the-digital-era/; or knowledge. insead.edu/responsibility/how-the-digital-economy-has-exacerbated-inequality-9726; (thanks to George Metakides for these references).

[3] There is even a new source of inequality that stems from bias in data, an issue which we won't discuss here further.

[4] As a result, look at the respective elections and the success of the populist right.

- *Ethical technology.* Being aware of the impact of our artifacts, we recognize the need to develop new technology along ethical guidelines. Informatics departments worldwide have included ethics in their curricula, either as stand-alone courses or embedded in specific technical subjects. Also, industry has come along; some companies even offer specific tools, and associations such as IEEE provide guidelines for ethical design of systems.[5] So, if we follow such guidelines, offer courses, and behave ethically, then it will work, at least in the long run. That's simple.

  But reality may again be a little bit complicated. Most of the research in AI, especially with respect to machine learning, is done by the big IT platform companies; they have the data and, with sufficient financial resources, also outstanding expertise. These companies try to anticipate "too much" regulation and argue for self-regulation. However, is this research really independent, is it not only "ethics washing" as observed by Wagner (2018)? Cases such as Google firing Timnit Gebru and Margaret Mitchell let the alarm bells ring.[6] But it is not only about independence of research, it is also about reproducibility of results, transparency of funding, or the governance structure of research (see Ebell et al. 2021). And there are other subtle problems, e.g., we argue for fairness in recommendation or search results. But how to define fairness, is it with respect to the provider of information or products, is it with respect to readers or consumers (and which sub-group), or do we need to define fairness with respect to some general societal criteria? One step further: let's assume these issues are solved and we all behave according to Kant's categorical imperative; can we then guarantee overall ethical behavior or a good outcome? Assuming the concept of human technology co-evolution, we have an evolutionary "optimization" process, which may lead to a local but not to a global optimum (e.g., preventing automatically monopolistic structures). Even more, this evolution "does not evolve on its own", but is—as in our context—governed by existing unequal societal and economic power relationships. So, ethics alone may not be enough.

- *It is about the economy.*[7] The digital transformation as a socioeconomic-technical process has to be put into a historical context. One could apply contemporary economic theory to understand what is going on (the "invisible hand" according to Adam Smith, i.e., following their self-interest consumers and firms create an efficient allocation of resources for the whole of society). However, the economic world has been substantially changed by the digital transformation. The value of labor is in the process of being reduced by increasing automatization and a possible unconditional basic income. These are simple observations but what are the implications? What is, ultimately, the role of humans in the production

---

[5] IEEE P7000 – IEEE Draft Model Process for Addressing Ethical Concerns During System Design. https://standards.ieee.org/project/7000.html

[6] https://www.bbc.com/news/technology-56135817

[7] We omit "stupid," for not offending the reader.

process? Or even, what is the value of a company? Can this still be captured and understood by traditional theories?

This is again complicated, as personal data seems to become the most distinguished feature people can contribute (at least on the Web) in this new world. Apparently, it is less and less the surplus ("Mehrwert") generated by humans in the labor process that is relevant but rather the added value by a never-ending stream of data, i.e., their behavioral traces on the Web. This data is used to develop, to train, and to optimize AI-driven services. Thus, users are permanently doing unpaid work. A user is, therefore, all three, a customer, a product, and a resource at the same time (Butollo and Nuss 2019). Furthermore, "instead of having a transparent market in which posted prices lead to value discovery, we have an opaque market in which consumers support Internet companies via, essentially, an invisible tax" (Vardi 2018). All this is related to the central role of online platforms. However, the investigation of their technological dominance and the resulting imbalances of power may require a network analytical perspective, integrating informatics, statistics, and political science. But such novel approaches to understand the new rules of the economic game and the mechanisms that drive the data-driven digital revolution are complicated. However, as far as we know there is no accepted method to measure the value of the data economy, or data itself. Data is the core of this development and is even called by several observers as the "gold nugget" of today. Whereas external valuation is difficult, the large online platforms are aware of the situation and are investing heavily. We need creative people with eyes from different disciplines in order to come up with further enlightening—both methodological as well as practical— insights. Remember Industrial Revolution once more: understanding the steam engine does not immediately result in Marx's *Critique of Political Economy*.

- *And about politics.* It is already everyday knowledge that Informatics will continue to bring about profound changes. All this seems to be an automatism, even like a force of nature. However, we neither think that there is a higher being that is responsible nor, in a similar mindset, that developments strictly follow a "historical determinism." If we, the people, should be the driving force, the simple approach would be that all people participate in decisions of shaping their own future, be it via democratic elections or via participatory initiatives.

However, in practice experiences are contradictory and, thus, complicated. An example: although it was social media that claimed to lay the basis for participative processes,[8] recent years have shown that their effect is often enough going in the opposite direction and fuels the loss of trust in policy makers (and thus in democracy in the long run). In addition, policy makers seem to be, at least sometimes, powerless against market automatisms, which in turn leads to the tendency to let people vote for "strong men." Today it is the platforms themselves that make inherent political decisions when, for instance, banning individuals or entire opinion groups. In conclusion, with the simplistic vision that the Internet

---

[8] In fact, social media contributed to broad political movements such as the Arab Spring.

will foster participation, we have ended up in a complex situation that shifted power from the people to global players which take opaque actions. And these quasi-monopolists not only run the "visible" global platforms but, in the background, have built up a critical infrastructure for the functioning of the entire economy (cloud service, machine learning services, etc.). Something needs to be done, that's the simple request. But available options are all difficult: e.g., building one's own (e.g., European) infrastructures, or regulations, or even a nationalization of these companies. Each of these alternatives raises complicated questions: should it take place on a global or national or regional scale? Regarding regulation, what should be addressed, the entire technology stack or just the software upper layer? Regarding nationalization, which is anyway a controversial act on its own, who would even have the power to do this? Or who should operate and control such infrastructures? There won't be simple answers ahead, but long and cumbersome discussions on a global scale, which are often devastatingly ineffective. As a blueprint, we mention the discussion on a Tobin-Tax as one lesson learned from the 2008 crisis and which did not materialize in any form after 13 years of debate.

- *We as academics.* Coming back to our role as academics, at a first glance it appears to be simple: specifically, in the technical disciplines we try to solve open questions, test new hypotheses, find novel models to describe the world, etc. Our papers are reviewed by colleagues and we present our results to the community in journals and conferences. And in some rare cases, these results find their way to the real world, making it hopefully and sometimes a better place.

  Clearly, life is not as simple as that; are we always aware of all effects of our contributions? Have we thought about our responsibility, as stated by Popper (1971)? Scientific technical solutions are not always for the better; sometimes such solutions and artifacts even worsen the situation. In addition, such solutions depend on the context and may reflect current societal structures and relationships. When normative claims come into play, things immediately become more complicated, and norms can be seen to conflict with the freedom of research. Especially in technology, scientists might be reluctant, as at least some often perceive their research activities as detached from social values and norms. Moreover, the COVID crisis has brought the ambiguous relation between politics and science and research on the agenda once more, demonstrating that we need to be aware of the pitfalls and consequences when determining this relation (Habermas 1970). We as researchers have to communicate that there are two sides of science, a "ready-made science" and a "science-in-the-making" (see Denning and Johnson 2021 or Latour 1987). Science is also a process, starting from struggling for what we need to know, towards an agreement on theories and models, i.e., settled science. Both sides are needed, as are—sometimes loud— debates between researchers. Thus, in certain situations we cannot provide 100% certain answers, and this must be communicated to the public, to politicians. A related aspect is the rapid development of technology, constantly (re-)shaping our world, which frightens people and increasingly leads to criticism of technology; sometimes it is hard to trust technological progress. As an example, as digital transformation leads to automatization and reorganization of work, certain jobs will disappear, which in turn raises an intrinsic reservation about modern technology (nonetheless, it has to be emphasized that it might also be a blessing for the society as a whole if certain jobs disappear). The complicated issue is how to manage this progress, with a societal or a market perspective, only.

Like other socioeconomic or sociotechnical-economic processes, digital transformation is a dialectic endeavor, full of contradictions, raising simple and complicated issues with often no easy solutions. One needs to understand technology to build a technical infrastructure for the human and the society, but at the same time this goal is only achievable if there is broad support from the people. Scientific insights need to be communicated and participation is required ("Citizen Science"). People have to understand the power they (still) have in this new society where fundamental rules are under change, but basic knowledge on these issues is often severely lacking. As an example, also the COVID crisis showed that privacy issues are often raised in weird ways: we have seen huge reservations against contact tracing apps, while the same people mark their vaccination appointments in their Facebook stories.

Digital Humanism is not only about fundamental and applied research, where different disciplines have to cooperate, but is also about different types of activities that have to be integrated, from research to innovation, education, policy briefing, and communication with the public. As simple or complicated as it may be, it needs democratic ways forward and solutions. Or to put it simply: at the end, techno-societal issues need political answers.

References

Butollo, F., and Nuss, S. (2019) Marx und die Roboter. Vernetzte Produktion, künstliche Intelligenz und lebendige Arbeit. Dietz Berlin (in German).

Danziger, S., Levav, J., & Avnaim-Pesso, L. (2011) Extraneous factors in judicial decisions. Proceedings of the National Academy of Sciences, 108(17), 6889-6892.

Denning, P., and Johnson, J. (2021) Science Is Not Another Opinion. Communications of the ACM. 3 (64)

Ebell, C., Baeza-Yates, R., Benjamins, R. et al. (2021) Towards intellectual freedom in an AI Ethics Global Community. AI Ethics (2021). https://doi.org/10.1007/s43681-021-00052-5

Habermas, J. (1970) "Technology and Science as 'Ideology.'" In Toward a Rational Society: Student Protest, Science, and Politics, trans. Jeremy J. Shapiro. Boston: Beacon Press. (Original article: Habermas, J. (1968) Technik und Wissenschaft als "Ideologie". Man and World 1 (4):483-523.)

Kahneman, D. (2011) Thinking, fast and slow. London: Penguin Books

Latour, B. (1987) Science in Action: How to Follow Scientists and Engineers through Society. Harvard University Press.

Meehl, P. E. (1986) Causes and effects of my disturbing little book. Journal of personality assessment, 50(3), 370-375.

Popper, K. R., (1971) The moral responsibility of the scientist. Bulletin of Peace Proposals, 2(3), 279-283

Vardi, M. (2018) How the hippies destroyed the Internet. Communications of the ACM. 7 (61).

Wagner, B. (2018) Ethics as an escape from regulation. From "ethics-washing" to ethics-shopping? In: Emre, B., Irina, B., Liisa, J.U.A. (Hg.): Being Profiled: Cogitas Ergo Sum. 10 Years of 'Profiling the European Citizen'. Amsterdam University Press, Amsterdam, pp. 84–88.

Werthner, H. et al. (2019) The Vienna Manifesto on Digital Humanism. https://dighum.ec.tuwien.ac.at/dighum-manifesto/

Correction to: Did You Find It on the Internet? Ethical Complexities of Search Engine Rankings

Cansu Canca

Correction to:
Chapter 19 in: H. Werthner et al. (eds.),
Perspectives on Digital Humanism,
https://doi.org/10.1007/978-3-030-86144-5_19

The original version of the chapter was inadvertently published with an error. The affiliation of the author Cansu Canca has now been corrected to "AI Ethics Lab, Cambridge, MA, USA".

The updated online version of this chapter can be found at
https://doi.org/10.1007/978-3-030-86144-5_19

Printed in the United States
by Baker & Taylor Publisher Services